SpringerBriefs in Plant Science

For further volumes:
http://www.springer.com/series/10080

Muhammad Asif

Progress and Opportunities of Doubled Haploid Production

 Springer

Muhammad Asif
University of Alberta
Edmonton, AB, Canada

ISSN 2192-1229 ISSN 2192-1210 (electronic)
ISBN 978-3-319-00731-1 ISBN 978-3-319-00732-8 (eBook)
DOI 10.1007/978-3-319-00732-8
Springer Cham Heidelberg New York Dordrecht London

Library of Congress Control Number: 2013941872

Printed on acid-free paper

Springer is part of Springer Science+Business Media (www.springer.com)

Foreword

The importance of haploidy or single totipotent cell to form a haploid plant is well known to plant scientists in the field of agriculture and related disciplines. The work on haploidy started in 1921 by A. D. Bergner who reported this fascinating phenomenon in *Datura stramonium* L. Since then, numerous findings have been reported in various crop species and the efforts to improve doubled haploid production resulted in the discovery of various methods like anther culture, isolated microspore culture, and wide hybridization. In crop plants, the ability to produce identical/homozygous individuals from single cells has dramatically reduced timescale to develop new cultivars and doubled haploidy has now become an essential biotechnology tool in plant breeding programs. The single cell-culture system provides many opportunities for process improvement, and genetically identical and physiologically uniform single cells are also being used as targets for cell biology, embryology, and genetic engineering studies.

On planning this monograph, my intent was to discuss the importance of haploidy in a variety of areas from fundamental to applied research and how molecular methods have been exploited recently to unravel/explore some of the underlying aspects of this fascinating developmental phenomenon of doubled haploids. Consequently, the brief is divided into six chapters. The introductory chapter (Chap. 1) provides information to the readers regarding history, production methods, and types of haploids. The next three chapters (2, 3, and 4) highlight various steps involved in the production of doubled haploids via androgenesis, gynogenesis, and parthenogenesis. The major bottlenecks of doubled haploid production like low frequency of green plant production and albinism have been discussed in detail along with major achievements that have changed the status of many recalcitrant crop species to responsive over the last 90 years. The use of doubled haploidy in plant breeding program is an effective strategy to achieve homozygosity in one generation and doubled haploid populations are being used extensively to map quantitative trait loci/genes of interest. Unicellular microspores and haploid embryos are main targets of mutation breeding and genetic transformation studies, as discussed in Chap. 5. Chapter 6 summarizes the brief along with future prospects of doubled haploid production.

I am indebted to Dr. Dean Spaner for his help and valuable suggestions in completing the brief. I am grateful to Dr. Rong-Cai Yang, Dr. Habibur Rahman, Dr. Harpinder Randhawa, and Dr. Francois Eudes for their constant support and encouragement. I thank all graduate students and research associates particularly Klaus Strenzke, Eric Amundsen, Atif Kamran, Leslie Bihari, Enid Perez, Cristiana Hill, and Muhammad Sajad for their help in various ways. The financial support of Canadian Wheat Board is greatly acknowledged.

Edmonton, AB, Canada   Muhammad Asif

Contents

Abbreviations

ABA	Abscisic Acid
$AgNO_3$	Silver Nitrate
APM	Amiprophos-methyl
ATP	Adenosine Triphosphate
BA	Benzyl Adenine
BAP	Benzylaminopurine
$CuSO_4$	Copper Sulfate
CWRS	Canada Western Red Spring
CWSWS	Canada Western Soft White Spring
DH	Doubled Haploid
DMSO	Dimethyl Sulfoxide
ELS	Embryo-Like Structures
FHB	Fusarium Head Blight
HSP	Heat Shock Proteins
IAA	Indole Acetic Acid
IMC	Isolated Microspore Culture
MCS	Multi Cellular Structures
MS	Murashige and Skoog
NAA	Naphthalene Acetic Acid
NADPH	Nicotinamide Adenine Dinucleotide Phosphate Hydrogen
NH_4NO_3	Ammonium Nitrate
PAA	Phenyl Acetic Acid
PCD	Program Cell Death
PEG	Polyethylene Glycol

QTL   Quantitative Trait Loci

SSD   Single Seed Descent

TDZ   Thidiazuron

$ZnSO_4$ Zinc Sulfate

Chapter 1
History, Production Methods, and Types of Haploids

Doubled haploidy is an efficient and effective research tool to obtain complete homozygosity within a heterozygous progeny in a single step. Since its discovery in 1921, various mechanisms/methods have been developed to produce doubled haploid plants, and the technique is constantly improving. Doubled haploidy has been adapted in plant breeding programs for many decades. It is a method of choice in crop species that are highly responsive and in which haploid production methodology/protocol has been well established. This chapter deals with the history of doubled haploids, production methods, and types of haploids.

1.1 Overview and History of Haploids

The haploid plants are considered sporophytes due to the presence of gametic chromosome number (n) in the cells. These plants are generally derived from male or female gametic cells. In monocots, doubled haploid plants can be produced from both male and female gametic cells but in dicot species, the available choice is only one cell. The haploids also occur in nature that develop/originate when egg cell or synergid directly develop into an embryo without the fusion of male and female gametes but these haploids are normally abnormal. Haploid plants have been developed in 100 species of angiosperms (Vasil 1997) and the phenomenon of haploidy is being practiced in many crop species like wheat (Inagaki 2003), maize (Gaillard et al. 1991), barley (Hagberg and Hagberg 1980), rice (Bishnoi et al. 2000), millet (Powell et al. 1975), sorghum (Brown 1943), oat (Nishiyama 1961), brassica (Thompson 1974), tomato (Kirillova and Bogdanova 1978), coca (Lanaud 1988), and cotton (Turcotte and Feasto 1974) for the production of haploids or doubled haploids. The formation of embryo from egg without involvement of sperm cells is called semigamy, and this type of haploidy has been observed in cotton where embryo formation takes place from the independent division of egg and/or sperm cells. This independent division of egg or sperm cell

M. Asif, *Progress and Opportunities of Doubled Haploid Production*, SpringerBriefs in Plant Science 6, DOI 10.1007/978-3-319-00732-8_1, © Springer International Publishing Switzerland 2013

(semigamy) is a heritable trait and is controlled by *Se* allele (Hodnett 2006). In laboratory conditions, embryogenesis is normally achieved by changing environmental conditions of anther or microspores by various means, which often results in the development of an embryo/multicellular structure as an alternative to pollen grain. In case of haploid production, the isolated microspore culture is particularly important to plant breeders/geneticists due to the presence of embryogenic microspores in large number that can develop into hundreds of doubled haploid plants under favorable conditions.

First overview of haploid production was given by Riley (1974) who reported that haploid production started in 1921 by A. D. Bergner who observed haploidy in *Datura stramonium* L. His research work was reported by Blakeslee et al. (1922). Since then, numerous findings of haploid production have been reported in various crops like tobacco (Clausen and Goodspeed 1924) and wheat (Gaines and Aase 1926). Due to the importance of haploid production, this phenomenon has increasingly motivated plant breeders/geneticists to investigate various methods of haploid production to be able to come up with one that can produce doubled haploids on a larger scale (Kimber and Riley 1963). These efforts have resulted in the discovery of various methods that include wide hybridization, parthenogenesis, alien cytoplasm, pollen irradiation, and sparse pollination (Kasha and Maluszynski 2003). The adoption of haploidy in maize breeding programs was initiated by Chase (1952). The maize haploid plants were produced by parthenogenesis that was followed by chromosome doubling to make them doubled haploids. The major breakthrough in this technique was achieved when Guha and Maheshwari (1964) successfully produced embryos from anthers of *Datura innoxia*. Bourgin and Nitsch (1967) followed the same procedure and produced haploids in *Nicotiana tabacum* and *N. sylvestris*. At present, haploid production methods such as isolated microspore culture, anther culture, wide hybridization, and ovule culture are among the methods of choice and they are employed in various crops to develop haploids. However, response to produce haploids differs greatly from species to species. Rapeseed, barley, and tobacco are considered as the most receptive crops to this technology, and these crops are often named as "model crops" but even in these crop species, haploid production response is mainly genotype dependent. The leguminous crop species are recalcitrant to haploid production. Cereals are highly genotype dependent, and the frequency of albino plants is much higher as compared to green plants when produced either through anther or isolated microspore culture (Holme et al. 1999). However, haploid production through wide hybridization has resolved the issue of albinism to some extent as "bulbosum method" is being used in barley to develop varieties on a larger scale. Similar advantages of wide hybridization has also been observed in other cereal species like macaroni wheat (Jauhar 2003), wheat (Inagaki 2003), oat (Rines 2003), and triticale (Wedzony 2003) by using maize as a male parent in the crossing. The major drawbacks of wide hybridization include labor intensive (due to enormous amount of emasculation, crossing and embryo rescue), time consuming, and the need for synchronization of male (pollen donor) and female plants of different species/genera involved. Other haploid production methods like anther culture or isolated microspore culture do not have these drawbacks.

1.2   In Vitro Haploid Production Methods

A variety of methods have been used to induce embryogenesis in male and female gametes to produce double haploids (Fig. 1.1). Maluszynski et al. (2003) comprehensively reviewed methods and protocols to produce haploids or doubled haploids

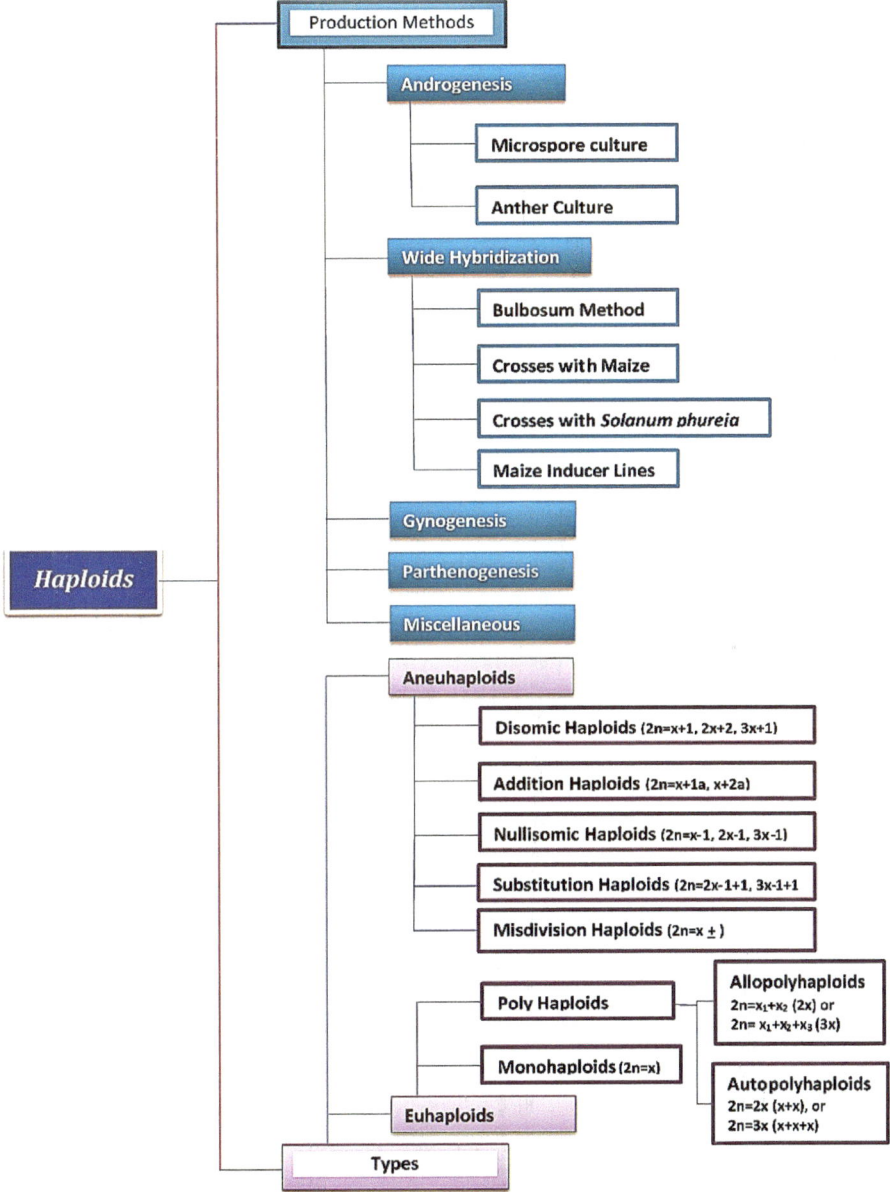

Fig. 1.1   Classification and production methods of haploids

in many crops. The methods that have been developed to induce embryogenesis are androgenesis (including anther culture and isolated microspore culture), gynogenesis, wide crossing, parthenogenesis, sparse pollination, pollen irradiation, centromere-mediated genome elimination, alien cytoplasm and seeds with twin embryos. In this book, the main focus has been given to four techniques/methods that have been widely used to produce doubled haploids on a larger and commercial scale.

- *Androgenesis*: It refers to the production of an embryo or zygote that carries chromosomes only from the male parent. In case of androgenesis, embryogenesis is induced in anthers or microspores directly or indirectly through callus formation.
- *Gynogenesis*: In this method, unfertilized ovary, or cell of embryo sac or ovule directly lead to the development of an embryo.
- *Wide Crossing*: In this method, two distantly related (outside of immediate gene pool) parents are crossed. The chromosomes of the pollinator parent are eliminated due to their nonhomology with those of female parent, and the resulting embryo contains chromosomes of the female parent only, which are subsequently doubled.
- *Parthenogenesis*: It is a type of asexual reproduction in which unfertilized egg cell develops into an embryo by semigamy, pseudogamy, or apogamy.

1.3   Types of Haploids

Haploidy refers to the numerical changes in chromosome number that can involve whole set of chromosomes (euhaploids) or only a part of it (aneuhaploids). The euhaploids will have half the number of chromosomes whether it is derived from a diploid or polyploid species. The changes in whole set of chromosomes will give rise to monohaploids ($2n=x$) and polyhaploids ($2n=2x$, $3x$, $4x$, $5x$, …). Therefore, polyhaploids can be dihaploids ($2n=2x$), trihaploid ($2n=3x$), tetrahaploid ($2n=4x$), pentahaploids ($2n=5x$), and so on. Polyhaploids derived from polyploid species can be further divided into autopolyhaploids or allopolyhaploids. Autopolyhaploids consists of multiple copies of the basic set of one particular genome (AAAA or BBBB) as in case of potato (*Solanum tuberosum* L., $2n=4x=48$), whereas allopolyhaploids have multiple copies of the basic set but from different genomes (ABD) as in case of wheat (*Triticum aestivum* L., $2n=6x=42$). In contrast to euhaploids, aneuhaploids may originate by either gain (called as hyperploidy) or loss (hypoploidy) of one or more chromosomes. If the gain in chromosome originate from basic set (x), the plants are called as disomic haploids ($2n=x+1$) but if gain in chromosome occur from alien species, the plants will be termed as addition haploids ($2n=x+1a$). The loss of one chromosome form gametic set will be termed as nullisomic haploid ($2n=x-1$). Aneuhaploids can also arise by substitution of one or more chromosomes (substitution haploids) by exact number from other or alien species ($2n=x-1+1$). The haploids which do not fit into the above-mentioned categories are termed as misdivision haploids. The classification of haploids (Fig. 1.1) has been extensively reviewed by Kimber and Riley (1963) and Gupta (2005).

References

Bishnoi U, Jain RK, Rohilla JS, Chowdhury VK, Gupta KR, Chowdhury JB (2000) Anther culture of recalcitrant indica×Basmati rice hybrids. Euphytica 114(2):93–101. doi:10.102 3/A:1003915331143

Blakeslee AF, Belling J, Farnham ME, Bergner AD (1922) A haploid mutant in the jimson weed *Datura stramonium*. Science 55(1433):646–647

Bourgin JP, Nitsch JP (1967) Production of haploids nicotiana from excised stamens. Annales De Physiologie Vegetale 9(4):377–382

Brown MS (1943) Haploid plants in sorghum. J Hered 34(6):163–166

Chase SS (1952) Production of homozygous diploids of maize from monoploids. Agron J 44(5):263–267

Clausen RE, Goodspeed TH (1924) Inheritance in *Nicotiana tabacum*. IV. The trisomic character, "enlarged". Genetics 9(2):181–197

Gaillard A, Vergne P, Beckert M (1991) Optimization of maize microspore isolation and culture conditions for reliable plant regeneration. Plant Cell Rep 10(2):55–58. doi:10.1007/BF00236456

Gaines EF, Aase HC (1926) A haploid wheat plant. Am J Bot 13(6):373–385

Guha S, Maheshwari SC (1964) In vitro production of embryos from anthers of datura. Nature 204(495):497. doi:10.1038/204497a0

Gupta PK (2005) Cytogenetics. Rastogi, Meerut

Hagberg A, Hagberg G (1980) High frequency of spontaneous haploids in the progeny of an induced mutation in barley. Hereditas 93(2):341–343. doi:10.1111/j.1601-5223.1980.tb01375.x

Hodnett GL (2006) The effect of the semigamy (se) mutant on the early development of cotton (*Gossypium barbadense* L.). Texas A&M University, College Station, TX

Holme IB, Olesen A, Hansen NJP, Andersen SB (1999) Anther and isolated microspore culture response of wheat lines from northwestern and eastern Europe. Plant Breed 118(2):111–117. doi:10.1046/j.1439-0523.1999.118002111.x

Inagaki MN (2003) Doubled haploid production in wheat through wide hybridization. In: Maluszynski M, Kasha KJ, Forster BP, Szarejko L (eds) Doubled haploid production in crop plants. Kluwer Academic, Dordrecht, pp 53–58

Jauhar PP (2003) Haploid and doubled haploid production in durum wheat by anther culture. In: Maluszynski M, Kasha KJ, Forster BP, Szarejko L (eds) Doubled haploid production in crop plants. Kluwer Academic, Dordrecht, pp 161–166

Kasha KJ, Maluszynski M (2003) Production of doubled haploids in crop plants. In: Maluszymski M, Kasha KJ, Forster BP, Szarejko I (eds) Dubled haploid production in crop plants. Kluwer Academic, Dordrecht, pp 1–4

Kimber G, Riley R (1963) Haploid angiosperms. Bot Rev 29(4):480–531. doi:10.1007/bf02860814

Kirillova GA, Bogdanova EN (1978) Comparative-study of haploid tomato form existing for a long-time and homozygous diploid form obtained from it. Genetika 14(6):1030–1037

Lanaud C (1988) Origin of haploids and semigamy in *Theobroma cacao l.* Euphytica 38(3): 221–228. doi:10.1007/bf00023524

Maluszynski M, Kasha KJ, Forster BP, Szarejko I (2003) Doubled haploid production in crop plants: a manual. Kluwer Academic, Dordrecht

Nishiyama I (1961) Cytogenetic studies in avena. 8. Haploid plant of sand oats (*Avena strigosa*). Jpn J Genet 36(3–4):72–75. doi:10.1266/jjg.36.72

Powell JB, Hanna WW, Burton GW (1975) Origin, cytology, and reproductive characteristics of haploids in pearl millet. Crop Sci 15(3):389–392

Riley R (1974) The status of haploid research. In: Kasha KJ (ed) Proceeding of the first international symposium on haploids in higher plants: advances and potential. University of Guelph, Guelph, pp 3–9

Rines HW (2003) Oat haploids from wide hybridization. In: Maluszymski M, Kasha KJ, Forster
    BP, Szarejko I (eds) Dubled haploid production in crop plants. Kluwer Academic, Dordrecht,
    pp 155–160
Thompson KF (1974) Homozygous diploid lines from naturally occurring haploids. Fette Seifen
    Anstrichmittel 76(7):303
Turcotte EL, Feasto CV (1974) Methods of producing haploids: semigametic production of cotton
    haploids. In: Kasha KJ (ed) Haploids in higher plants. Advances and potential. II. Methods of
    producing haploids. University of Guelph, Guelph, pp 53–64
Vasil IK (1997) In vitro haploid production in higher plants. Kluwer Academic, Dordrecht
Wedzony M (2003) Protocol for anther culture in hexaploid triticale (*Triticosecale Wittm.*). In:
    Maluszynski M, Kasha KJ, Forster BP, Szarejko L (eds) Doubled haploid production in crop
    plants. Kluwer Academic, Dordrecht, pp 123–128

Chapter 2
Androgenesis: A Fascinating Doubled Haploid Production Process

Androgenesis is one of the most important methods that have been extensively used in plant breeding programs to produce double haploids. It involves the induction of microspore embryogenesis that leads to the development of a haploid embryo instead of mature pollen grain. The microspore embryogenesis is usually brought about by modifying the environmental conditions of anthers/microspores by reprogramming their gametophytic pathway towards sporophytic growth and development. Under natural conditions, the microspore develops into a mature pollen grain that comprises of generative and vegetative nuclei. The generative nucleus develops into two sperm nuclei. Thus, the sporophytic development should be started before the onset of cell division when the gamete cells in the microspores are still totipotent. However, the embryogenic stage of microspores varies greatly among species (Touraev et al. 2001). The microspores are amenable to androgenesis and consist of haploid (n) number of chromosomes and therefore; give rise to haploid plants. Androgenesis can be divided into three distinguished steps (1) embryogenesis induction (2) regeneration of haploids followed by (3) artificial chromosome doubling. The production of haploids or doubled haploids (DH) via androgenesis can be achieved either through isolated microspore culture or anther culture.

2.1 Microspore Culture

Microspore culture (pollen culture) offers an opportunity to the plant breeders to develop DH plants on a larger scale which enables them to speed up the breeding process by fixing homozygosity in one generation after a cross has been made. Thus, cultivar development period can be dramatically reduced with the help of isolated microspore culture (IMC) in crop species responsive to this method. The IMC involves isolation of immature pollens or microspores from the anthers

M. Asif, *Progress and Opportunities of Doubled Haploid Production*,
SpringerBriefs in Plant Science 6, DOI 10.1007/978-3-319-00732-8_2,
© Springer International Publishing Switzerland 2013

followed by culturing them on growth media under optimum environmental conditions necessary for their growth and development and to reprogramme their gametophytic pathway towards sporophytic by using various kinds of stress treatments. This method is preferred over anther culture (Kieffer et al. 1993; Arnison and Keller 1990) due to the following reasons:

- Anther culture can give rise to diploid plantlets (non-haploids) from anther tissues (wall) along with haploids, whereas in microspore culture plantlets always originate from microspores that have haploid number of chromosomes (n).
- Anther culture is a lengthy, time consuming, and laborious method.
- In case of anther culture, anther tissues other than the microspores could have a destructive influence on the growth and development of a developing microspore and to some extent, it deters their developmental process.
- The developing microspores have uniform nutrient accessibility during IMC.
- Isolated microspore culture provides an opportunity to better track the pathway of microspore embryogenesis by monitoring each embryogenesis stage separately and to better understand the most important factors contributing towards microspore embryogenesis.
- Isolated microspore culture offers a platform for targeted mutagenesis and an effective gene transfer technique that can direct the breeder/molecular biologists to pyramid genes of their interests in a shorter period of time. Moreover, the transgenic plants can be identified at a very early stage of their life cycle.
- Cell changes studies during a shift from gametophytic to sporophytic pathway and the initiation of microspore embryogenesis can be easily performed/tracked during IMC.
- The embryogenic units are ten times greater in IMC as compared to anther culture.

Microspore embryogenesis gained enormous importance and attention from the breeders after 1960. This technique developed very rapidly when Guha and Maheshwari (1964) discovered that by providing specific environmental conditions, haploid plants could be easily produced from the anthers of immature pollen grains of *Datura innoxia*. Nitsch (1974) used natural shedding phenomenon and successfully isolated microspores of *Nicotiana* sp. from its anthers that followed mechanical microspores isolation in *Brassica* sp. by Lichter (1982) before culturing them on media to produce haploid plants. This discovery opened a major breakthrough in the area of DH production. Since 1960, extensive research studies have been conducted to improve the efficiency of IMC. However, a large number of crop species such as legumes are still recalcitrant to microspore culture. Each step of IMC has been investigated in detail to improve the efficiency of this technology. A wide range of protocols have been summarized in various crops. Each crop/genotype has its own specific protocol due to distinct androgenic response to microspore embryogenesis but main processes involved in this technique are same (Fig. 2.1) that include (1) donor plant's growth and developmental conditions, (2) removal/collection of floral organs from donor plants, (3) pretreatments, (4) isolation of microspores, (5) composition of media, (6) regeneration of haploids and (7) artificial chromosome doubling.

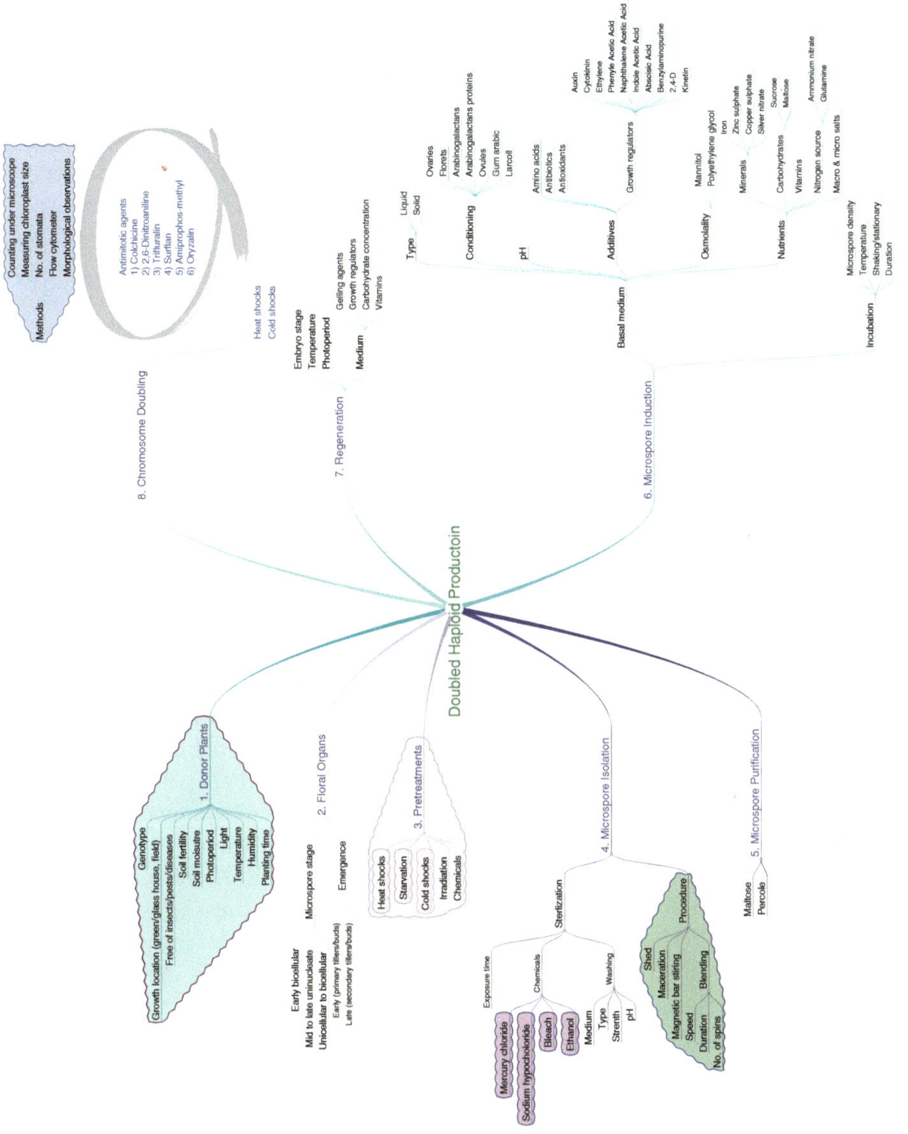

Fig. 2.1 Steps involved in IMC to develop haploid/doubled haploid plants

2.1.1 Donor Plant's Growth and Developmental Conditions

The donor plant's growth and developmental conditions occupy the most important position in the whole process of IMC as efficiency is directly linked to it. If the donor plant is free of insects, pests, diseases, absence of nutrient and water deficiencies and environmental stresses like temperature, humidity, and photoperiod, the effectiveness of this method can be enhanced to a greater extent.

The donor plants are normally grown in optimum conditions to get a healthy crop stand e.g., in cereals, strong and vigorous tillers are desirable. Other agronomic practices like watering and fertilization are done routinely. An improved embryogenic response has been observed if donor plants are planted under controlled conditions (green house, glass house or control chambers) than plants grown under field conditions. Optimum growth and developmental conditions can be provided to donor plants through supply of optimum light, humidity, temperature, and photoperiod under controlled conditions that also ensures to minimize disease occurrence and infestation due to insects and pests. The growth chamber grown donor plants are often preferred over green house plants because higher number of green plants can be obtained from them (Dahleen 1999). Field grown plants have also been used to isolate microspores but less embryogenic response has been observed. Moreover, there are greater chances of contamination if donor plants are grown under field conditions. The growing conditions of donor plants have a direct effect not only on the embryogenesis but also on the regeneration and the number of green plants. Among growth conditions, temperature is of major concern that has been probed in several studies. Luk et al. (1983) reported that an increase in day and night temperature from 18 to 28 °C and 14 to 25 °C, respectively, in triticale will diminish the regeneration process to a greater extent. The plants sown in cold temperature consequently give rise to higher number of embryos and green plants (Bernard 1977). In this study, cold temperature (12–15 °C) gave better response in inducing embryogenesis in triticale. On the contrary, no such requirement of cold temperature treatment for donor plants of pepper and asparagus has been reported (Lantos et al. 2009; Wolyn and Nichols 2003). If plants are not grown under cold temperature, the stress to isolated anthers in the form of cold temperature has proven a strong positive effect on embryogenesis (Osolnik et al. 1993). Thus, on the basis of these studies it can be deduced that cold treatment is not only essential to arrest the gametophytic stage but it also helps in improving the entire androgenesis and regeneration developmental processes.

Light intensity and its interaction with temperature also affect the physiological status of donor plants. A five time decrease in embryogenic response in *Nicotiana tabacum* was noticed with increase in photoperiod from 8 to 16 h (Duncan and Heberle 1976). The status of nitrogen in soil also has a direct relationship with embryogenic response in tobacco and it has been observed that donor plants grown under starved nitrogen conditions have given enhanced response to IMC as compared to donor plants that were fertilized routinely (Tsay 1982). However, specific recommendations with regards to optimal conditions for the growth and development of donor plants are not possible because donor plant requirements vary significantly among various crop species.

The genotype of the donor plants also plays a crucial role in the response of microspores to embryogenesis and it not only differs from species to species but also varies considerably within species and this is especially true for cereals like triticale, barley, wheat, and oat. This intra and inter specific variation in embryogenic response during IMC differs extensively with some varieties/cultivars/lines of a particular genotype/species exhibiting a greater response while others showing no response at all, and sometimes differences in microspore embryogenesis within a plant are also very high (Phippen and Ockendon 1990). The winter and spring genotypes have given varying degree of response to embryogenesis. In *B. napus*, greater embryogenic response was noted in winter cultivars than in spring cultivars (Keller et al. 1987a, b). Contradictory results were obtained by Ohkawa et al. (1987) in 96 genotypes. Similarly, japonica genotypes in rice are more responsive to androgenesis than indica cultivars (Miah et al. 1985) and a same trend between *B. napus* and *B. juncea* has been reported by Chanana et al. (2005) where the latter carry a poor response. In wheat, 32–85.6 % of genotypic variation for embryogenic response was observed (Zhou 1996) and these differences were 73 % in barley (Torp et al. 2001). The embryogenic response of genotypes is considered a heritable character and embryogenic response can be improved by crossing a non or poor receptive cultivar with a well responsive one (Petolino et al. 1988). A number of experiments have been conducted to investigate the androgenic response of cultivars/lines/varieties in different species so that model species with improved overall response of DH production can be identified. The major varieties that have been recognized for their better androgenic response are Chris, Pavon 79 and Bob White in wheat (Kasha et al. 2003b), Igri in winter barley (Davies 2003), Topas in *B. napus* (Ferrie et al. 1995a), Narayen, Rupali, Kimberley in chickpea (Croser et al. 2011), CV-2 in *B. rapa* (Ferrie et al. 1995a), Green, Shogun, SDB9 in *B. oleracea* (Dias 2001) and CAV-2648 in wild species of red oat (Kiviharju et al. 2004).

2.1.2 Collection of Floral Organs

The efficiency of microspore culture is also dependent on plant age and pollen stage at which the floral organs are collected from donor plants for microspore isolation. A greater androgenic response has been noticed if microspore isolation is done with the floral organs that emerge first than those appearing in the later life cycle of donor plants. A similar trend have been seen in cereal species where primary tillers have given a much better response to anther as well as microspore culture than secondary and tertiary tillers. However, in *B. rapa* and *B. napus*, pollen collected or microspores isolated from older plants perform well to the androgenesis than the young donor plants (Takahata et al. 1991). In a sowing date study on *B. juncea*, it was observed that frequency of embryos was increased when floral organs were collected from late sown plants than the plants planted at normal sowing date (Agarwal and Bhojwani 1993). In tobacco plants, a four time variation in the number of green plantlets was reported with a variation of 2 mm corolla length (Dunwell 1976).

The optimum microspore stage that can reprogramme the microspores from game-tophytic to sporophytic pathway appears to differ among species. In case of *N. taba-cum*, first pollen grain mitosis or unicellular to bicellular (G_1) stage of microspores is considered to be the most responsive (Touraev et al. 2001) but on the other hand microspores between mid-late uni-nucleate or early bicellular are most responsive in cereals. The microspores isolated at later stages of pollen grain are normally not responsive because they contain starch grains (Sangwan and Sangwannorreel 1987a, b), whereas in brassica pollen grains used for microspore isolation already contain starch grains and embryogenesis is successfully induced in them by a well-timed heat pretreatment (Binarova et al. 1997). The DAPI and Acetocarmine stains have been extensively employed in tissue culture studies to identify the accurate microspore stage prior to their collection or before using them for isolation to induce embryogenesis (Fan et al. 1988).

2.1.3   *Pretreatments*

The pretreatments of various types are given to the floral organs to induce stresses that can ultimately help to switch gametophytic pathway of microspore to sporo-phytic development (Fig. 2.2). The commonly used pretreatments are cold and heat shocks, starvations in the form of nitrogen and carbohydrates, irradiation or chemi-cal treatments that are often given to the floral organs like spikes or floret buds, excised anthers or even to the microspores after their isolation. Pechan and Keller (1989) pointed out that pollen irradiation is not a widely adapted stress treatment as compared to other pretreatments. There are few species that do not require any pre-treatment (or stress) for embryogenic induction because their microspores exhibit certain kind of natural capability for microspore embryogenesis (Zhou et al. 1991). However, the frequency of such tendency towards embryogenesis is exceptionally low. It is also assumed that the removal of anthers or floral organs from the donor plant is itself a substantial amount of stress that can guide the fate of microspore towards sporophytic development. The pretreatments alone or in combination with each other act as triggering factors or as an external stimulus to achieve an optimum conversion of microspores from their gametophytic growth to sporophytic pathway. It was also observed that animal cells in addition to plant cells also require some sort of pretreatment in the form of stress to induce embryogenesis because origin of "Dolly," in case of animal cloning, entails stress pretreatment as one of the major component for development of an embryogenic cell in sheep (Zheng 2003). Puddephat et al. (1999) found that the donor plant developmental conditions in onion had a strong positive influence in inducing microspore embryogenesis. The pretreatments in the form of cold temperature have shown promising androgenic response in barley (Li and Devaux 2003), wheat (Indrianto et al. 1999), durum wheat (Sibi et al. 2001) and rice (Bishnoi et al. 2000) but on the other hand heat treatment enhanced embryogenic response in brassica (Binarova et al. 1997), tobacco (Touraev et al. 1996b), cucumbers (Gemes-Juhasz et al. 2002), pepper

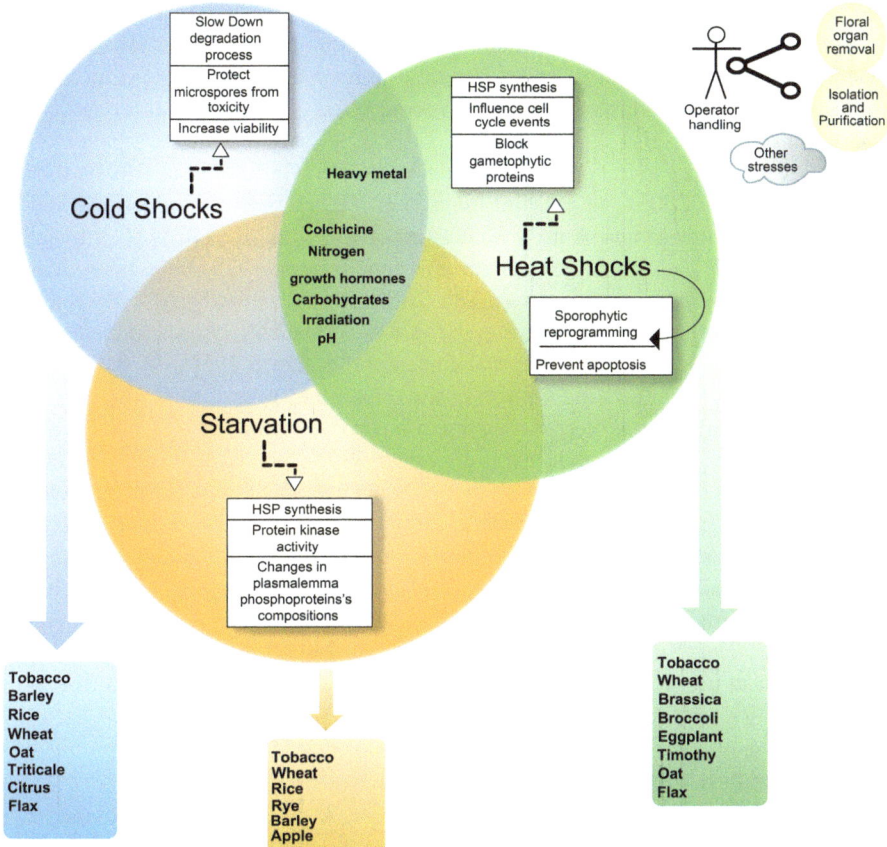

Fig. 2.2 Schematic representation of pretreatments/stresses employed during IMC in various crops

(Barany et al. 2001), wheat (Touraev et al. 1996b) and starvations in the form of nitrogen and carbohydrates conferred improved effect in tobacco (Touraev et al. 1996a), barley (Hoekstra et al. 1992) and rice (Raina and Irfan 1998). The use of colchicine and auxin as a pre-treatment has also induced microspore embryogenesis in few species (Obert and Barnabas 2004). The use of growth regulators such as abscisic acid, cytokinin, and auxin have been employed to switch somatic cell towards embryogenic cell (Filonova et al. 2000) but their transitional response in regards to DH production via IMC is not adequate/sufficient.

2.1.3.1 Cold Pretreatment

Cold pretreatment has been used in many crops to induce microspore embryogenesis. The cold pretreatment of anthers has a nursing effect on microspores that not only arrests their normal gametophytic development (Zheng 2003) but also assists to synchronize the whole developmental progression of microspores (Hu and Kasha 1999). The cold treatment of spikes for more than 7 days in cereals stimulates the microspore embryogenesis and helps to increase the frequency of embryos or multicellular structures and green plants. A substantial progress in embryogenic efficiency of microspores in cereal crops such as maize (Gaillard et al. 1991), triticale (Marciniak et al. 2003), wheat (Indrianto et al. 1999), barley (Sunderland and Xu 1982), rye (Immonen and Anttila 2000) and other crops like citrus (Germana and Chiancone 2003) and tobacco (Sunderland and Roberts 1979) have been done by applying cold pretreatment as stress. The chilling temperature helps to decrease the degradation of microspores/cells, thereby inhibiting their exposures to the decaying material and other toxic substances (Duncan and Heberle 1976). In *B. rapa*, cold temperature treatment helps to arrest the bicellular microspore stage, thereby increasing the frequency of embryogenic microspores exhibiting two equal nuclei. Sato et al. (2002) reported that bicellular microspore stage with two equal nuclei is one of the most crucial phase to induce embryogenesis in *B. rapa*. A cold treatment of 2–4 days has improved the effectiveness of microspore culture to several folds in *B. napus* whereas it is less helpful in *B. rapa* and has no effect in *B. oleracea* (Gu et al. 2003a, b; Xu et al. 2007). The cold pretreatment of spikes in barley helped microspore separation from anthers and their free occurrence in the locule (Sunderland and Xu 1982) whereas it does not have any effect on microspore detachment in tobacco because they are previously separated and found free in the locule and does not need any pretreatment (Zoriniants et al. 2005).

2.1.3.2 Heat Pretreatment

The heat pretreatment has been found as an efficient embryogenesis inducer. It has been used alone or in combination with sugar starvations to achieve maximum output from microspore embryogenesis and has been extensively applied in brassica species not only to induce embryogenesis but also to increase the frequency of embryo or embryo like structure (ELS) and green plants during IMC. When the floret buds of *B. napus* and *B. carinata* were treated with heat shocks for 1–4 days at 32 °C, it activated the process of embryogenesis in all buds used in the experiment (Pechan and Smykal 2001). However, the heat pretreatment longer than 4 days did not show any improvement in the cell division and lowered down the frequency/number of embryos to a greater extent (Barro and Martin 1999). The heat pretreatment (33 °C for 8 h) and treatment of floret buds exhibiting microspores at bicellular phase/stage at 42 °C in rapeseed showed better results for inducing embryogenesis. A gentle heat treatment (33 °C) of rice anthers also resulted in obtaining optimum numbers of embryos and green plants (Raina and Irfan 1998). In tobacco, mild heat shocks as a separate pretreatment or with sugar starvation helped to seize gametophytic development and improved microspore's reprogramming towards sporophytic pathway (Touraev et al. 1996c). It has been observed that heat shocks cause numerous cell modifications/alterations and of these changes, the synthesis/production of highly conserved group of heat shock proteins (HSP) carries an imperative position with respect to androgenesis. The production of HSPs is not only linked with heat treatments, but they also originate as a result of various motives such as osmotic stress, cold treatment, and oxidative stress (Almoguera and Jordano 1992; Sabehat et al. 1998). The HSPs synthesis can take place at different phases of plant growth like embryogenesis, fruit maturation (Low et al. 2000), pollen grain growth and development (Parcellier et al. 2003), and germination (Wehmeyer et al. 1996). Major components of these HSPs, mainly HSP90 and HSP70, are known to produce elevated level of expression right from the initiation of microspore embryogenesis or soon after embryogenic induction in *N. tabacum* (Zarsky et al. 1995), *B. napus* (Segui-Simarro et al. 2003), corn (Gagliardi et al. 1995), *Capsicum annuum* (Barany et al. 2001) and this HSP synthesis continues till the first pollen mitosis. These HSPs also obstruct the synthesis of those proteins that are needed for pollen grain growth and development. Thus, HSPs play a very crucial role to program the microspores from their gametophytic development to sporophytic pathway (Telmer et al. 1993). These HSPs are also known to play a major role to cope with program cell death (PCD) of microspores or microspore's apoptosis throughout their induction or culture period when they are subjected to heat shocks (Zoriniants et al. 2005).

2.1.3.3 Starvation

The sugar and nitrogen are major components with respect to stresses induced by starvation. The starvation stress is always applied to the uniform population of immature pollens or microspores mostly during the induction phase when the microspores are between mid to late uni-nucleate phase to induce sporophytic development. In tobacco, the heat shocks along with sugar and nitrogen starvation was given to a homogenous population of microspores between mid to late uninucleate and early bicellular stage that resulted in the induction of microspore embryogenesis in greater than 70 % of the microspore. The remaining 30 % microspores either died or did not possess/exhibit embryogenic characteristics that can only be attributed to the stress application and/or complications in the isolation process (Touraev et al. 1996c). Caredda et al. (2000) conducted a similar experiment by applying 3–4 days cold shocks instead of heat treatment in conjunction with starvation that resulted in enhanced survival rate of microspores to a considerable degree and increased the percentage of green versus albino plantlets compared to a treatment where floral organs of donor plants were pretreated only with cold for 3–4 weeks. The significant improvements in barley and wheat microspore cultures were reported by replacing sucrose by maltose in the induction medium that elevated the rate of metabolism leading to hypoxia and also resulted in increasing ethanol accumulation and lowering energy levels (Indrianto et al. 1999; Scott et al. 1995). Thus, the substitution slowed down the consumption of maltose that gave rise to a starvation stress in the microspores, directing them to initiate sporophytic development rather than gametophytic. The utilization of carbohydrates during microspore culture is mainly dependent on the pH and osmotic pressure of the media.

The induction of microspores in a media exhibiting low pH levels i.e., between 5.0 and 6.0 can result in an effective utilization of sucrose and facilitate the conversion process of sucrose into starch, thereby leading microspores towards pollen grain development following gametophytic pathway. On the other hand, if microspores are inducted in a media having pH levels 8.0 or higher, it will considerably reduce the sucrose utilization (creating a carbohydrate starvation) and direct the microspores towards sporophytic pathway (Zoriniants et al. 2005). Furthermore, a very slight variation in sugar contents of induction medium was reported by Zhou et al. (1991), suggesting their important role as an osmoticum in the media. These findings were followed by various research studies that reported an enhanced effect of medium osmotic pressure on embryogenic development of microspores (Croser et al. 2006; Ramirez et al. 2001). The sugar starvation initiated microspore's dedifferentiation, leading them towards sporophytic pathway but on the other hand when all necessary/required ingredients/nutrients were present in the induction medium, it resulted in the redifferentiation of isolated microspores, directing them towards normal gametophytic development (Harada et al. 1988). The nitrogen starvation induced by glutamine has shown a major role to suppress or inhibit the process of microspores maturation and enhanced their successful transfer to sporophytic development (Kyo and Harada 1986). The starvation induced by carbohydrates is known to cause various structural and physiological cell modifications that comprised of (1) inhibition of cell growth, (2) instant/speedy carbohydrate's intake, (3) reduction in the rate of cell respiration, (4) degradation of cell proteins and lipids, (5) rapid accumulation of free amino acids and Pi, phosphorylholine and (6) reduction in the activities of glycolytic enzymes (Yu 1999). Principally, these cellular modifications take place during the entire process of cellular adaptation to carbohydrate starvations. The cellular changes brought about by sugar starvation in tobacco microspores include dedifferentiation of plastids, degradation of starch and lamellar structure, emergence of large vacuole, dilution of generative cell wall, rapid decline in the size of nucleolus, loss of nuclear pore in the vegetative nuclei and various chromatin changes, that have been experienced when tobacco microspores from early and mid to bicellular phase are cultured to induce embryogenesis (Garrido et al. 1995; Kyo and Harada 1990). The other cellular changes associated with sugar starvation consisted of deregulation in protein kinase activities, decrease in energy levels especially in the form of ATPs, decrease in RNA synthesis along with status/levels of cell/microspore's respiration and especially these physiological and structural changes occurred when sucrose was substituted with maltose in the induction media (Scott et al. 1995; Zarsky et al. 1990). Moreover, it is also believed that cell cycle arrest along with the activation of HSP gene during tobacco embryogenesis as a consequence of stress treatment in the form of sugar starvation not only help to preclude PCD of microspores but also increase cell division as compared to cell enlargement (Zarsky et al. 1992, 1995).

Based on above discussion, it can be recapitulated that many crop plants require stress pretreatment either in the form of cold, heat, or carbohydrates to switch their gametophytic development of microspores to sporophytic pathway whereas on the other hand the crops that do not require such pretreatment, the physical removal of floral buds/spikes/organs from the donor plants produce/generate sufficient stress that can initiate androgenesis in microspores. However, the length of stress plays a significant role to induce microspore embryogenesis and to attain high percentage of embryos or ELS along with green plants but the accurate motive of how these stresses influence the degree or pace of androgenesis and regeneration process is still unclear. However, it can be contemplated that these pretreatments facilitate the overall process of microspore embryogenesis or trigger embryogenesis by creating various stresses that bring structural and physiological cellular changes leading them towards sporophytic event rather than normal gametophytic pathway of pollen grain development.

2.1.4 Microspore's Isolation and Purification

The floral organs such as buds, florets, and spikes are treated with chemicals prior to microspore isolation in order to eliminate/remove any insect, pest, fungal, or bacterial contaminants. The main chemicals used for removal of contaminants are bleach, ethanol, sodium hypochlorite, and mercury chloride. However, care must be taken with respect to duration of surface sterilization to avoid any lethal effect of these chemicals on the microspores. The surface sterilization procedure starts by immersing floral organs in 75 % ethanol or 10 % bleach for 3–5 min. The floral organs can also be surface sterilized with 6 % sodium hypochlorite for 15–20 min. These surface sterilization procedures are then followed by water (doubled distilled) washings (2–3) for about 1–2 min. In few studies, the surface sterilization with mercury chloride (0.1 %) has been conducted but it is recommended to avoid it for surface sterilization of floral organs that are going to be used for isolation of micro-spores due to its lethal or toxic effect.

Four microspore isolation procedures have been reported in various studies that include shed microspore, maceration, magnetic bar stirring, and blending. The shed microspore method was first reported in *N. tabacum* by Sunderland and Roberts (1977). It comprised of microspores shedding from anthers in liquid media that was followed by their induction in a different media separate from anthers to circumvent toxic effect of anther (somatic) tissues. The anther tissue is critical to remove and it is recommended to keep these tissues away from microspores because they release phenolic compounds that have a lethal effect on microspores. Moreover, somatic tissues of anthers may lead to development of diploid (2*n*) rather than haploid plant-lets and may give rise to some complications in the research experiments/trials. This isolation method is very simple, easy to follow, avoid any complications for isolation and always results in less injury/damage to the microspores but it is more like anther culture rather than microspore culture. Since the discovery of shed microspore in 1977, it was quickly adopted by Datta and Wenzel (1987) for microspore isolation in wheat. The magnetic bar stirring isolation method involves a stirring force to remove microspores that are covered by anthers. This isolation procedure is more efficient and effective as compared to natural microspore shedding because it gives higher number of microspores than the shed microspore method (Cho and Zapata 1990). Lichter (1982) used glass or Teflon rod to isolate microspores from anthers in *B. napus* by pestle maceration followed by sieving for purification. Micro blending is one of the most widely adopted methods that consist of blending the surface sterilized dissected buds or florets in mechanical blenders. It removes microspores more effectively as compared to previously described procedures. The somatic tissues of anthers are effectively removed from microspores by sieving through sterile mesh of various sizes (100–200 μm). The mechanical micro blending was first described by Swanson et al. (1987) in *B. napus*, followed by Olsen (1991) in barley and Mejza et al. (1993) in wheat. Currently, a couple of centrifugations are conducted for microspore purification. These centrifugations often involves density gradients such as maltose (Kasha et al. 2001) or percol (Joersbo et al. 1990) to isolate microspores that are between mid to late uninucleate or early bicellular phase/stage in case of cereals and between unicellular to mid-bicellular phase in case of tobacco. This purification results in obtaining uniform population of isolated microspores that do not contain any anther tissue and nonviable microspores. The mechanical micro blending is considerably important procedure as compared to other methods as it always yield more number of viable microspores (75 %) (Gustafson et al. 1995). Lately, one more isolation procedure has been identified in *Datura metel* by Iqbal and Wijesekara (2007) where the anthers were aseptically removed from their filaments by opening the flower buds. Then, these anthers were used to isolate microspores by applying various combinations of temperature pulses. Anthers were placed lengthwise on liquid media and squeezed out the microspores by temperature pulse followed by removal of anther tissues or debris by using stereo microscope.

2.1.5 Media Composition

The basal media like MN6, MS, B5, A2, MMS3, P4, P2, CHB3, NLN, N6, and NPB99 has been effectively used in anther and microspores culture in many crop species. The NLN (Lichter 1982) and MS (Murashige and Skoog 1962) with minor changes are used for brassica and other allied species, whereas NPB99 (Konzak et al. 1999), A2 (Touraev et al. 1996b), and MMS3 (Hu and Kasha 1997) are routinely used during anther or microspore culture in cereal species like wheat, barley, and triticale. In early days of androgenesis, solid media using agar as a solidifying agent was preferred but as the time progressed, liquid media became a best choice to achieve desired results because solid media contains agar that is proven to have pollen inhibitory effect in few cases and hinder pollen growth towards embryogenesis. On the other hand, liquid media offers no competition for nutrient availability among developing embryos or ELS, especially during initial induction/culture phase of anthers/microspores. The main problem associated with liquid media is microspore sinking that often results in creating an anaerobic environment leading towards slower metabolism and decrease in energy production. However, this problem can be easily solved by adding Ficoll in the media (Cistue et al. 2009; Kao

1981). The effectiveness of IMC is mainly dependent on the properties and characteristics of induction medium that consists of (1) nutrient constituents like mineral substances, carbohydrate, pH, and osmolality; (2) cultural environment like light intensity, temperature, photoperiod, and duration of culture; and (3) density of the medium. The role of media with respect to microspore culture is twofolds: first, it supplies microspores with all necessary nutrients required for their growth and development in in vitro and secondly, it also helps to switch their pathway from gametophytic to sporophytic. It has been recommended that microspores must be provided with all required nutrients rich medium having macro and micro salts, carbohydrates, vitamins, nitrogen source, and growth regulators, if required. The nutrient concentration and their presence in the media are highly variable and depend on the crop species being used.

For quite some time, hormones such as potato extract, auxins, coconut milk, cytokinins, yeast extract, and ethylene were frequently used in the media (Raghavan 1986) but recently it has been reported that these growth hormones have a major role in callus formation during the process of embryogenesis. The characteristics and functions of growth regulators/hormones in media have been investigated in detail to see their effect on increasing the efficiency of embryogenesis in many crop species. In cereals like barley, triticale, and wheat, the positive role of Phenyl Acetic Acid (PAA), Naphthalene Acetic Acid (NAA), Indole Acetic Acid (IAA), Abscisic Acid (ABA), Benzylaminopurine (BAP), 2,4-D, and Kinetin in the media alone or in combination with each other to improve entire process of embryogenesis have been reported in numerous research studies (Davies 2003; Hansen 2000; Kasha et al. 2001; Otani and Shimada 1994; Pauk et al. 2003). However, these hormones have not been used to a larger extent in media being used for microspore culture. Antibiotics such as cefotaxime have been successfully used in the microspore induction medium to manage/tackle the problem of contamination (Davies 2003; Lantos et al. 2006). Charcoal has also been added to media to control contamination or to remove toxin substances due to its absorption capacity but simultaneously it also absorbs other crucial nutrients from the media necessary for the development of ELS (Gland et al. 1988). The addition of antioxidants in media to promote embryogenesis has also been examined where glutathione played an important role in embryo development (Asif et al. 2013; Fletcher et al. 1998). Sucrose is a key source of carbon in media. The concentration of sucrose differs from one crop species to another. Carbohydrates are only source of energy but their role as an osmoticum to maintain a certain osmotic pressure in media cannot be overlooked and as an osmoticum, they regulate the movement of nutrients/elements from cells.

The media alteration is one the most popular exercise that has been carried out for the last 40–50 years to seek maximum output from microspore culture for improving androgenic response especially in recalcitrant species. Substitution of sucrose by maltose between 60 and 90 g/l in rice, triticale, barley, rye, and wheat induction media (Bishnoi et al. 2000; Chu et al. 1990; Karsai and Bedo 1997; Kasha et al. 2003a; Otani and Shimada 1994; Pauk et al. 2000) have demonstrated improved effects but on the other hand sucrose is still being used in Brassica species in a concentration of 130 g/l (Pechan and Smykal 2001). A significant improvement in the efficiency of barley embryogenesis has been illustrated by altering the sources of organic nitrogen. Olsen (1987) observed enhanced results by lowering down the concentration of ammonium nitrate (NH_4NO_3) and increasing glutamine concentration in media. This finding is still being adopted by researchers and recently glutamine in the concentration of 500 mg/l in media has revealed a positive influence on barley microspore culture (Kasha and Maluszynski 2003; Kasha et al. 2003a, b). Glutamine is also a key element to develop DHs in brassica (Hansen 2003), rye (Pulli and Guo 2003), and triticale (Wedzony 2003) via microspore culture. In durum wheat, glutamine in combination with glutathione has given promising results with respect to the frequency of embryos and green plants (Cistue et al. 2009; Asif et al. 2013). The mineral ingredients like Fe, $ZnSO_4$, and $CuSO_4$, were also reported to have a positive effect in inducing embryogenesis and increasing ratio of green vs. albino plants in *Hordium vulgare* (Echavarri et al. 2008; Wojnarowiez et al. 2002). Jacquard et al. (2009) and Prem et al. (2008) reported a positive effect of induction medium supplemented with Cu and $AgNO_3$. In a similar manner, the supplementation of media with n-butanol also improved embryo yield in wheat microspore culture and boosted the frequency of green plants up to 3–5 times (Soriano et al. 2008). In tobacco, embryogenic division of microspores is highly reliant on presence of Fe in the induction media. Furthermore, Fe also plays a major role in the senescence of anther wall (Vagera and Havranek 1983).

The pH and osmotic pressure of media are other important factors that play a critical role in affecting not only the process of embryogenesis but also help in improving regeneration efficiency of embryos towards green plants. The alteration of medium osmoticum is usually done using polyethylene glycol (PEG) and mannitol in different concentrations. The occurrence of albinism has been seen to be lessened by high osmolality of the medium (Jacquard et al. 2006). The media pH is normally kept around 6.0, however, slight change is needed depending on crop species being used for embryogenesis (Ferrie et al. 1995b).

The IMC in cereals often comprises of induction medium supplemented with various types of embryogenic material such as florets, ovaries, embryogenic microspores, or ovules that has shown promising results in improving an overall process of microspore embryogenesis in wheat, barley, triticale, and rye (Lantos et al. 2009). Generally, it is assumed that supplementing induction media with these tissues supply microspores with certain phytohormones and signaling molecules to start/initiate the process of embryogenesis and thus, these tissues contribute towards embryo development but main function of this material/tissue in converting microspores to embryos is yet not clear. However, it was pointed out that arabinogalactans-proteins/ arabinogalactans exhibit certain stimulatory functions that helped to initiate the process of embryogenesis (Letarte et al. 2006). The supplemented induction media with gum arabic and Larcoll had also shown strong impact in improving wheat microspore culture. The addition of Larcoll with or without ovaries in the induction medium greatly reduced mortality of microspores. It also provided a genotypic independent effect, reduced albinism, and improved green plants regeneration (Letarte et al. 2006). In barley microspore culture, the addition of florets in the induction medium had greatly improved androgenesis and had been found more successful than ovary co-culture (Lu et al. 2008).

The optimum density of microspores in the induction media is another factor that ensures their further growth and development and decides the time required to produce embryos from microspores. In triticale, the microspores density of 3×10^4 to 2×10^5 microspores/ml of the induction media is considered optimum for normal growth and development of microspores (Eudes and Chugh 2009) while densities of $8–10 \times 10^4$ microspores/ml for *Capsicum annum* (Kim et al. 2008), 4×10^4 microspores/ml for *B. napus* (Huang et al. 1990), and 5×10^4 microspores/ml for *B. oleracea* (Ferrie et al. 1999) are considered ideal.

2.1.6 Regeneration

The development of embryos from microspores can be achieved via indirect/direct pathways. Regardless of these pathways, embryos are required to shift from culture to regeneration medium in order to achieve a smooth transition/switch from embryos/ELS to green plants. Regeneration is always achieved in the presence of light. This transition relies on many factors such as age or growth stage of embryos at the time of regeneration, regeneration media, light intensity, and temperature during regeneration period/phase. The cold treatment below 10 °C during early regeneration period (1–3 weeks) and transfer of embryos at cotyledonary stage in brassica species has shown promising results (Ferrie 2003; Niu et al. 1999). However, the transition phase of microspores descended embryos (torpedo, early, mid or late cotyledonary stage) differs greatly from one species to another. The desiccation of embryos prior to their transfer to regeneration medium was recommended by Hansen (2003) who pointed out that desiccation facilitates the process of embryo germination. The cold temperature around 10 °C for 7 days during initial phase of regeneration has also been suggested for triticale to alleviate the process of transition, reducing albinism, and enhance the frequency of green plants (Wedzony 2003). The supplementing regeneration media with vitamins or phytohormones and drought stress to the embryoids has also been recommended to ease the overall transition from embryos to green plants (Zhang et al. 2006).

2.1.7 Increase in Ploidy Level

The number of chromosomes/ploidy level of plantlets produced by IMC can be verified/determined by several methods that include (1) counting of chromosomes that is mostly done using microspore (2) by measuring chloroplast size and number of stomata (guard cells) (3) using flow cytometer and (4) through morphological observations. A comparison of these methods was done by Sari et al. (1999) to determine ploidy level in water melon. The authors concluded that stomatal or guard cell measurement is one of the simplest and easiest methods to find/calculate number of chromosomes in plants. Spontaneous chromosome doubling is high in few crop species; thus, these species do not require any artificial chromosome doubling whereas on the other hand, most crop species need an increase in their ploidy level to covert haploid plants (obtained through IMC or anther culture) to doubled haploids that often involve the use of anti-microtubule agents like colchicine.

The rate of chromosome doubling is affected by numerous factors such as type of genotype, stage of microspores at the time of floral organ collection from donor plants, pathway of microspore during embryogenesis, pretreatment (cold, heat, or starvation), exposure time of microspores/embryos to various chemical agents, concentration of chemicals during induction and regeneration phases, and methods of application. The history of induced doubling of chromosomes goes back to 1929 when Lindstorm (1929) decapitated tomato shoot and discovered that new developing shoots were tetraploid rather than diploid. This was followed by Randolph (1932) who conducted an experiment in maize and induced an artificial increase in ploidy level by giving heat treatment (using heating pot) to the developing ear shoot. This artificial chromosome doubling attracted many researchers to design studies/experiments in order to test different methodologies/protocols to induce chromosome doubling in agricultural crops using cold treatments, heat shocks, and antimitotic agents. The discovery of colchicine from Gloriosa by Clewer et al. (1915) perfected its use in chromosome doubling and with the passage of time, it became a method of choice for artificial chromosome doubling (Blakeslee 1939) in at least 48 agricultural crops. Regardless of its extensive exploitation, effectiveness, and application in agricultural crops, there are numerous disadvantages associated with its usage that include occurrence of mixed polyploids (Pei 1985), loss of sterility, decrease in fertility, abnormal growth, chromosomes rearrangements, and gene mutations (Luckett 1989). However, it has been successfully used for chromosome doubling to produce/develop DH plants in wheat, sorghum, barley, maize, sugar beet, and many other crops.

Principally, the doubling of chromosome is achieved by various means/pathways that include (1) endomitosis that is referred as "duplication of chromosome number without nuclear division," (2) an interference in cell cycle of plants (3) endoreduplication referred as "chromatids become double without separating from each other" (4) C-mitosis referred as "an artificially induced abortive nuclear division where separation of centromere does not take place in the metaphase stage," (5) nuclear fusion where "one nucleus forms as a result of fusion of two or more nuclei" (Jensen 1974). The entire cell cycle can be divided into four well-defined stages (Francis 2007): (1) G_1 (Gap$_1$): A post mitotic stage in which cell grows and enlarges and it becomes ready for cell division, (2) S (Synthesis): It is characterized by DNA replication or synthesis, (3) G_2 (Gap$_2$): A pre-mitotic stage, and (4) M (Mitosis): that consists of division of a mother cell into two daughter cells. The chemical agents that interfere with cell division at the completion of Synthesis/S stage and prior to the completion of Mitosis/M has been termed as good agents to increase ploidy level in plants (Dhooghe et al. 2011). Various pretreatments that have been applied to donor plants or microspores to initiate their sporophytic growth have revealed explicit results with respect to increase in ploidy levels e.g. pretreatment of floral organs with mannitol alone or in combination with cold or heat shocks have significantly increased the frequency of chromosome doubling in wheat (Li and Devaux 2003) and rye (Guo and Pulli 2000b). On the other hand, colchicine treatment alone or in combination with heat or cold shocks in the induction medium has also facilitated chromosome doubling in *Phleum pretense* (Guo and Pulli 2000a), *B. napus* (Zhao et al. 1996a, b), and Easter lily (Antoine and Beckert 1997). Few other anti-microtubule agents have also been successfully exploited for this purpose that include 2,6-Dinitroaniline in watermelon, Trifluralin in Orange Ball Buddleia, Surflan in *Lilium longiflorum*, amiprophos-methyl (APM) in *Dianthus* sp., and oryzalin in *Solanum* sp. (Greplova et al. 2009; Nimura et al. 2006; Omran et al. 2008; Takamura et al. 2002; Van 2008). These anti-microtubule agents induce chromosome doubling by creating hindrance in the separation/segregation of sister chromatids toward poles, inhibit spindle formation and nuclear fusion (Testillano et al. 2004), and offer extreme affinity to plant tubulins as compared to the most commonly used colchicine. Thus, a very little amount of these anti-microtubule agents (mostly in millimole concentration) is required to induce artificial chromosome doubling (Morejohn and Fosket 1984).

On the basis of above discussion, it can be concluded that a universal protocol cannot be identified or developed for artificial chromosome doubling mainly due to the complexity of the process and its genotypic dependency because anti-microtubule agents behave differently in different crops. Therefore, selection of a polyploidizing agent is mainly dependent upon the type of genotype being used, stage of cells/cell cycle at the time of application, application procedure/method, and exposure time to these chemicals. Nevertheless, colchicine is an extensively used and widely adapted anti-mitotic agent to induce chromosome doubling in cereal, leguminous, and horticultural crops but the applicator should avoid any contact during its treatment/use mainly due to its anticipated lethal and harmful effects on plants and to applicator as well.

2.1.8 Albinism

In plants, albinism can be defined as lack or deficiency of green pigment called "chlorophyll" or failure to carry out the process of "photosynthesis," a chemical process necessary to synthesize food (carbohydrate) from carbon dioxide (CO_2) and water (H_2O) in the presence of sunlight. Albinism eventually results in plant death. The process of photosynthesis is initiated by absorbing light energy by round, oval, or disc shaped structures/organelles known as chloroplast which consists of chlorophyll. Plants store absorbed light in the form of Nicotinamide Adenine Dinucleotide Phosphate Hydrogen (NADPH) and Adenosine Triphosphate (ATP). This captured light is then used by plants in later stages. Green pigment or chloroplast organelles are absent in albino plants and therefore, they are not able to carry out photosynthesis, a process essential for their growth and development. Thus, these plants do not reach maturity and die at a very early stage. Albinism is considered a major bottleneck in plant genetics and breeding programs that involve interspecific crosses or wide hybridization to create variation and in plant tissue culture techniques involving microspore and anther culture particularly in case of cereals such as barley, oat, wheat, rice, rye, and triticale. Varying degree of albinism have been described in DH production via anther culture, microspore culture, and wide hybridization that is characterized by a partial to total loss of green pigments. Abadie et al. (2006) conducted an interesting experiment to compare chlorophyll contents of green and albino plants and depicted severe dissimilarities. They reported chlorophyll contents of green and albino plants as $2.97+0.56$ and $1.9+5 \times 10^{-2}$ µg/mg of fresh weight, respectively. Yao and Cohen (2000) also pointed that albino plants have at least 1–6 % less green pigments than green plants. In wheat, the importance of magnetic field was studied by Pingping et al. (2011) to improve an overall chlorophyll contents in the leaves of chlorophyll deficient plants. They reported an increase in the chlorophyll content of magnetically treated albino plants, which converted them to partially green plants and such plants attained physiological maturity as well. A similar study carried out in date palm to investigate the influence of magnetic field on chlorophyll contents and results showed a substantial improvement in total pigment contents, carotenoid, and chlorophyll a, b due to static magnetic field. However, the pigment content increase was extremely reliant on exposure, duration and intensity (Dhawi and Al-Khayri 2008). A similar trend in the increase of chlorophyll contents in soybean after an exposure to static magnetic field has also been reported by Atak et al. (2007). Mouritzen and Holm (1994) depicted that during earlier stages of plant growth, albino plants can be distinguished from green plants because of variations in their plastid DNA as a result of microspores/anthers redifferentiation and such plants have irregular chloroplast shape/structures rather than the normal that can be differentiated either by amyloplasts or proplastids (undeveloped). Therefore, these plants are not able to carry out the process of photosynthesis and cannot make carbohydrates for their growth and development. Caredda et al. (2000) described that seedlings/plantlets of albino plants can exploit reserve food only for some time and when stored carbohydrates are exhausted, albino plants begin to die due to reason that abnormal and undeveloped plastids cannot be switched to the functional chloroplasts. The dissimilarities with respect to physiology, structure, and behavior of plastids in albino and green plants have also been reported in barley (Caredda et al. 2000, 2004). These findings revealed that genotypes yielding/giving green plants exhibit thylakoids (dense and undifferentiated plastids capable of multiplying quickly and accumulating starch rapidly) while albino plants were devoid of thylakoids, deficient in cytoplasm, plastids were not dividing and starch was accumulating in their stroma. Furthermore, plastids in genotypes producing green plants were having much higher levels of DNA as compared to other genotypes.

The decrease in frequency of albino plants has been achieved by manipulating genetics (using different cultivars) as well as by altering growing conditions of donor plants. Various aspects like growth and developmental conditions of donor plants, genotype, stage of microspores at the time of floral organs collection, pretreatments, induction duration, microspore pathway, composition of medium, embryo age at the time of transfer to regeneration medium, temperature, oxygen, and light intensity during induction and regeneration play a crucial role in tackling this challenge. The alteration in any of these components will alter the frequency of green to albino plants.

The cold shocks and starvation for 3–4 days have significantly decreased the frequency of albino plants as compared to a longer pretreatment of 3–4 weeks

(Kasha et al. 2001). In an exciting study in barley, supplementation of induction media with $CuSO_4$ alone or in combination with mannitol enhanced the percentage (90 %) of green plants as compared to chlorophyll deficient plants in a cultivar Igri (Cistué et al. 2003). It has been suggested that a strong relationship exist between microspore sampling stage and frequency of chlorophyll deficient plants (Caredda and Clement 1999). The frequency of albino plants increases if the donor plants of barley and oat are grown in a temperature less than 15 °C (Collins 1927). However, an exposure of oat albino mutants to a temperature higher than 20 °C results in switching to green plants (Nishiyam and Motoyosh 1966). As discussed earlier, chlorophyll deficient plants/plantlets are not capable of making their own food or enough carbohydrates to keep and support their growth and development, therefore, it has been proposed that raising sugar (especially sucrose) levels/contents of the medium can facilitate to solve the problem of albino plants. Saidi et al. (1997) successfully tackled this issue (albinism) in *Triticum turgidum* by manipulating the sucrose contents of media and converted albino plants/plantlets to green. The addition of starch-melibiose and mannitol in medium in combination with cold treatment considerably enhanced green plant frequency in barley (Datta and Potrykus 1998; Hunter 1987). The green pigment content and percentage of green plants have also been significantly increased by the addition of glucose and growth hormones such as cytokinin, kinetin, IAA, and benzyl adenine (BA) (Broughton 2008; Chory et al. 1991; Nishiyama and Motoyoshi 1966) but kinetin and 2,4-D seemed to have no influence on the green plant percentage especially in triticale (Pauk et al. 2000). In anther culture or IMC, addition of Ficoll in the liquid induction medium has prevented microspore's sinking that in turn help to decrease DNA degradation in plastids. Therefore, the addition of Ficoll in induction medium in barley microspore culture enhanced the percentage (0 to 50 %) of green plants (Kao et al. 1991). The same trend of maltose (Redha and Talaat 2008) and Ficoll (Zhou et al. 1992) has been reported in wheat. An interesting aspect in regards to albinism has been highlighted in cereals (oat, barley, wheat, and rice) with respect to the collection of floral organs from donor plants (Reinbothe et al. 2003a, b) where the authors stated that collection of floral organs from primary tillers give higher number of green plants as compared to a collection from secondary or tertiary tillers that might be due to a hormonal imbalance in later (secondary or tertiary) tillers that affected the structure and behavior of plastids. Moreover, the existence of competition among secondary or tertiary tillers for hormones, and nutrients are concentrated in center of root zone might be a cause of albinism in later tillers (Casimiro et al. 2003) and it is also obvious because primary tillers are always more productive and healthy than later ones.

2.1.8.1 Genetics and Genomics of Albinism and Green Plant Regeneration

A large number of studies have been conducted to identify quantitative trait loci (QTL) and genes in several crop plants to increase the percentage of green plants. In this regards, QTLs on chromosomes IBL/1RS, 2AL, 2BL, and 5BL have been mapped to improve the frequency of green plants in wheat. Among them, QTL mapped on 2AL explicated higher variation (Torp et al. 2001; Tuvesson et al. 1989). In another experiment, two genes controlling embryogenesis have been identified on chromosomes 2D and 2A using wheat monosomic lines while 5B, 5A, 4A, and 2B carry few minor genes (Zhang and Li 1984). Genes controlling frequency of albino plants and embryoids have also been mapped on 5B and 5BL respectively (Agache et al. 1989). Recently, three QTLs were identified on barley chromosomes 6H, 5H, and 2H for number and percentage of green plants (Chen et al. 2007). It was previously reported that albinism is controlled by one gene in barley (Collins 1927). Two QTLs have also been identified on rice chromosomes 1 and 9 as well (He et al. 1998) to improve green plant percentage. Ekiz and Konzak (1991) conducted studies in wheat using alloplasmic lines exhibiting different plastid but same nuclear genome and depicted that plastid genome play a critical role in microspore culture response in wheat. The phenomenon of albinism is extremely heritable (Larsen et al. 1991) in some crops while in others such as wheat (Redha and Talaat 2008), low heritability has been reported for green plant percentage. However, Chaudhary et al. (2003) argued that it is strongly influenced by non-additive and additive type of gene action. In a similar study conducted by Moieni and Sarrafi (1995), on 49 different wheat varieties, high heritabilities ranging from 0.80 to 0.88 were found for characteristics such as green plant frequency, embryoid percentage, and frequency of plantlets regeneration. The specific and general combining abilities of these characteristics have also been found to be significant. The investigators demonstrated lowest heritability for albinism proposing that proportion of albino plants can be decreased by altering environmental and cultural conditions at various stages during the process of microspore culture in wheat. Few other experiments on this aspect highlighted that frequency of albino plants is under the control of one gene in soybean, barley, and maize (Barwale and Widholm 1987; Collins 1927; Neuffer et al. 1997) while two or more loci are involved in peanut (Dwivedi et al. 1984). In peanut, cytoplasm inheritance has also been shown to control albinism (Branch and Kvien 1992).

2.1.9   Pathways of Microspore Embryogenesis

The embryogenic capability in microspores is usually attained by the application of different stresses and/or starvation that is followed by a series of steps during which the microspores are converted to embryo/embryo like structures. During the process of embryogenesis, the microspore goes through several morphological, physiological, and cytological changes that are indispensable for their further growth and development in in-vitro. This embryogenic process of microspores can be distinguished into three distinct phases (1) attainment of embryogenic capability, (2) several asymmetric or symmetric cell divisions within exine wall that convert/lead microspores to embryo or ELS or sometime called multicellular structures (MCS) and (3) development and initiation of a certain pattern in ELS/MCS following exine wall disruption. Indrianto et al. (2001) carried out an interesting experiment to track microspores in wheat during the entire process of embryogenesis. They reported that microspores isolated from stress induced anthers were twofold increased in size and morphologically diverse as compared to microspores derived from freshly isolated anthers. On the basis of microspore morphology during embryogenesis, they divided microspores into three distinct kinds: Type I was characterized as vacuolated microspores because they were exhibiting a huge vacuole that was present in the center while their nucleus was pushed/pressed near periphery/cell wall. Basically, these types of microspores were non-stressed that are usually present at late unicellular stage. Type II was characterized as having fragmented/broken vacuole. Basically, cytoplasmic strands were crossing/passing through vacuole from one end to another and these strands were linked to phragmosome (cytoplasmic pocket) inside the nucleus. In this type (II) of microspores, cytoplasmic pocket was adjacent to cell wall while type III microspores can be easily differentiated from others because they were exhibiting a phragmosome in the center. Their study also revealed that 62 % of the type III microspores were not embryogenic as they were not capable to switch their gametophytic pathway to sporophytic and could not develop into embryos. On the contrary, this frequency was very less in microspores of type I (5 %) and II (23 %). Few other morphological and cytological variations in microspores during embryogenesis including cell enlargement, intine (new cell wall) development inside exine, chromatin compaction and decrease in size, amount, and magnitude of starch grains and nucleolus have also been reported (Garrido et al. 1995; Ramirez et al. 2001).

There are numerous studies that have discussed the pathway and fate of microspores regarding their switching or reprogramming from gametophytic pathway to sporophytic. Theories presented in these experiments are basically established on the basis of division of generative and vegetative cells. Recently, five potential pathways for conversion of embryogenic microspore to embryo or ELS have been purposed by Aionesei et al. (2005) that are basically a modification or alteration in A, B, and C pathways previously defined by Sunderland (1974). A-Pathway is characterized by a symmetrical division of vegetative cells leading towards embryo development. However, sometime callus development has also been reported instead of embryo formation and this is true especially for cereals like rice (Chen 1977), barley (Sun 1978), and wheat (Wang et al. 1973). In this type of pathway, generative cell dies at a very initial stage of embryogenesis or goes through a division that results in two sperm cells that die ultimately. The A-pathway has been noticed in the embryogenesis of *Brassica napus* (Fan et al. 1988), wheat (Reynolds 1993), and tobacco (Sunderla and Wicks 1971). B-Pathway is associated with the splitting of cell nucleus in two identical vegetative cells, both of them contribute in switching towards embryo formation as identified by Indrianto et al. (2001) in wheat. C-Pathway comprised of a merging or fusion of one vegetative and one generative cell nucleus or fusion between two vegetative cell nuclei. This Pathway has been reported in barley (Yao et al. 1997) and *Datura innoxia* (Sunderland 1974). D-Pathway is an altered type of B-Pathway whereby two broken/divided nuclei divide again and again that ultimately lead to production of callus or a haploid embryo (Pan et al. 1983; Zhu et al. 1978). E-Pathway consists of development of an embryo from generative nucleus. The repeated division of generative and vegetative nuclei can also give rise to production of an embryo but this embryo will consist of higher number of cells from generative than vegetative nuclei (Sun 1978). The E-Pathway has been noticed in wheat, rice, barley, and *Hyoscyamus niger* (Pan et al. 1983; Qu and Chen 1984; Raghavan 1976, 1978).

The preponderance of one pathway over other is governed by several elements. In this regard, pretreatments in the form of cold, heat, and starvation or any kind of stress play a critical role to decide the fate of microspores (Kasha et al. 2001). The adaptation of pathway varies considerably among species. Recent advancements in video cell tracking system and development of flow cytometry will definitely assist the molecular scientists to improve their understanding on how complex mechanism of microspore conversion from gametophytic to sporophytic pathway take place and ultimately culminate these microspores to haploid plants.

2.2   Anther Culture

The haploids plants through anther culture are usually obtained via two methods i.e., culturing anthers in liquid or semi liquid media that involves pollen separation by agitation, and placing anthers on solid media (solidification is usually obtained using agar). Basically, surface sterilized buds and florets are opened (in vitro) in sterile environment followed by anther removal and placing them on liquid or solid medium (Sunderland et al. 1984). Once the embryo formation is completed, embryos (culture) are shifted to the regeneration medium under light conditions for organogenic differentiation (shoot and root development). This method is very similar to IMC. The only difference in microspore culture is the removal of anther tissues or anther wall (somatic tissue) to prevent any lethal effect of maternal tissues on embryo development; and sometimes, somatic tissue may give rise to a diploid plant rather than haploid. Similar to IMC, anther culture also implies numerous pretreatments (vary considerably from species to species), surface sterilization, anther dissection, and finally placing them on induction and regeneration medium. The effectiveness of anther culture is highly reliant on growth, developmental conditions and physiological state of donor plants, pollen or microspore stage at the time of anther dissection, genotype, and media composition. In anther culture, the pollen may give rise to callus (indirect embryogenesis) tissues or callus formation as in wheat and rice or lead to an embryo (direct embryogenesis) development as in *Brassica* sp.

2.2.1   Genotype, Physiological State, Growth and Developmental Stage of Donor Plants

Physiological state, growth and developmental conditions and genotype of donor plants are among the important factors that decide efficiency of anther culture because these conditions directly interfere with overall effectiveness of embryogenic pollen grains (P-grains) by effecting hormonal level and nutritional status of anther tissues (Sunderland and Dunwell 1977). It has been reported that donor plants grown in nitrogen starvation conditions often yield embryogenic pollen grains. These pollen grains can be easily differentiated by having a large vacuole, absence of starch grains, and presence of a thin exine wall (Heberle-Bors 1984, 1989). Contrary to IMC, better response has been obtained from field grown donor plants as compared to green house plants (Vasil 1980). The other developmental conditions such as day light, photoperiod intensity, and temperature also influence anther culture to a considerable degree (Heberle-Bors 1989). A varied genetic response has also been noticed in several experiments/studies that differ not only among species but also within genus, species, and cultivars thereby suggesting its major role in anther culture/embryogenesis. Germana (2007) and Bajaj (1980) conducted two different studies to identify most responsive cultivar to anther culture. They termed 2 out of 23 and 10 out of 20 varieties as responsive in citrus and wheat, respectively. In anther culture, the embryogenic time/window of pollen grain consists/starts from the first mitosis which is characterized by vacuolated microspores to bi-cellular. However, this embryogenic window is highly variable and depends on the genotype being used for anther culture. Moreover, pollen grain loses their embryogenic efficiency when they begin storing/preserving starch in the form of grains (Raghavan 1990; Touraev et al. 2001). The ploidy level of plants obtained through anther culture is also influenced by the developmental stage of pollen grains/microspores at the time of induction. Sopory and Munshi (1997) depicted that microspores at uninucleate stage will give rise to haploid plants while culturing of anther having microspores at later stages often yield higher ploidy levels.

2.2.2 Pretreatments and Media Composition

The pretreatments have been categorized as "novel," "widely used," and "neglected." The commonly adapted pretreatments are cold and heat shocks, starvation in the form of nitrogen and sucrose, heavy metal and chemical treatments, changes in pH, humidity, osmotic levels, and water stresses (Shariatpanahi et al. 2006). Among these pretreatments, temperature shocks have been termed as widely adapted. As discussed earlier, the anthers are usually chosen when they exhibit microspores in the embryogenic window (between first mitosis to bi-cellular phase), however, a little heat (41 °C) pretreatment in B. napus have resulted in acquiring embryogenic division in already developed vegetative cells as reported by Binarova et al. (1997) and it also appeared to be helpful in pepper (Barany et al. 2001) and Nicotiana tabacum (Touraev et al. 1996a, c) anther culture. The gamma rays have been successfully used as pretreatments for anther culture in barley (Vagera et al. 2004) and rice (Aldemita and Zapata 1991) while colchicine has also been used as pretreatment (stress inducer) as well as to double the chromosome number in various crops such as wheat, brassica, maize, rice, sugar beet, and sorghum (Germana 2011a, b).

The composition of media also occupies an important position in anther culture to induce embryogenesis. In this regard, B5, MS, and N6 with minor changes are among the widely adapted media. The MS is mostly used in solanaceous crops while N6 has been applied in cereals (Chu 1981). Sucrose is a main source of carbohydrates and its concentration varies from 6 to 17 %. In anther culture, medium having high sucrose concentration have been used in species where culturing of tricellular pollen (mature) has given high response (e.g. in cruciferae) (Dunwell and Thurling 1985) but on the other hand in solanaceous species where bicellular pollen is used for anther culture, medium with low concentration of sucrose has given optimum results (Dunwell 2010). Maltose in the medium has also indicated explicit influence on anther culture embryogenesis in rye, rice, triticale, barley, and wheat (Wedzony et al. 2009). Germana and Chiancone (2003) described explicit findings of clementine anther culture using galactose and lactose in medium. The same results in clementine have also been reported using sucrose in combination with glycerol (Germana et al. 2000). The effects of growth hormones have been extensively studied during the last 60–70 years and they have provided exceptional results in some recalcitrant species. However, there are few species (belonging to solanaceae) that do not need any growth hormones in their culture for embryogenesis. There are two main functions of growth hormones in the medium i.e., one is to induce embryogenesis (Bajaj 1990; Bajaj et al. 1977) and other is to identify the fate of embryogenic pathway (Ball et al. 1993). Various studies have indicated that 2,4-D help to enhance/promote callus growth while NAA and IAA supports direct embryogenesis (Liang et al. 1987). The supplementation of medium with polyamines has also improved the frequency of embryos in clementine (Chiancone et al. 2006), *Cucumis sativus* (Kumar et al. 2004), and wheat (Rajyalakshmi et al. 1995). Similar findings in cereal anther culture have also been stated using arabinogalactans (Letarte et al. 2006) and ovary co-culturing (Broughton 2008).

An incredible advancement in anther culture methodology has been achieved in crop species such as triticale, wheat, rice, barley, rye, and several others like medicinal, vegetables, fruits, ornamental, and woody plants. However, there are still many group of species that are termed as recalcitrant to anther culture and legumes are considered as one of them (Dunwell 2010; Wedzony et al. 2009). The up to date progress and developments of anther culture have been recently reviewed in detail by Dunwell (2010), Germana (2011a, b), Touraev et al. (2009) and Wedzony et al. (2009).

## 2.3	Uniparental Chromosome Removal/Elimination or Wide Hybridization

The uniparental chromosome elimination or wide hybridization is considered to be an important tool not only to produce DH but also to create genetic variation, introduce new species, and for gene transformation studies. It consists of crossing a female parent to a distant male exhibiting haploid inducer genes. Intercrossed parents are taxonomically or ecologically similar to each other. During the process of intercrossing, chromosomes of pollen donor parent are automatically removed or eliminated. Wide hybridization becomes a best method to achieve desired results in DH production following the recovery of barley (*Hordeum vulgare*) haploid plants involving wide intercrossing using *H. bulbosum* as a male parent (Kasha and Kao

1970). During wide hybridization, endosperm is either not developed or poorly formed. Thus, embryo must be rescued or cultured in-vitro that otherwise may not survive and give rise to a haploid plant. The in-vitro embryo culture provides a conducive environment and nurtures the immature or weak embryos allowing them to carry their growth and developmental process. Cereals such as wheat, barley, rice, maize, rye, and triticale are amongst the most privileged crops in which wide hybridization have been exploited extensively along with microspore and anther culture to induce sporophytic embryogenesis. The technique of wide hybridization has been effectively used in solanaceous crop species to recover hybrids. The key benefits of wide hybridization include absence of gametoclonal variation, genotypic independence, getting unbiased random gametes for producing mapping populations, and absence of albino plants or albinism which is especially true for cereals.

2.3.1   Bulbosum Method

The intercrossing of *H. vulgare* and *H. bulbosum* involves preferential chromosome removal of later parent following fertilization. The completely grown embryos (caryopsis) are rescued (in-vitro) before endosperm disintegration, usually 12–14 days after pollination, to recover hybrids or haploid plants. The chromosome elimination is genetically controlled and genes involved in the chromosome elimination of *H. bulbosum* chromosomes have been mapped on *H. vulgare* chromosomes 2 and 3 (Ho and Kasha 1975). It has been further explained that elimination or retention of chromosomes is highly genotypic dependent (Pickering 1984) and it will only take place if the parents are grown in a cold temperature below 18 °C together with the application/spray of growth hormones/regulators (like 2,4-D, or Dicamba) after 1–2 days of pollination as illustrated by Devaux and Pickering (2005). Chromosomal elimination can be tracked by distinct arrangement of species specific centromeres on multi polar spindles along with the production of nuclear extrusions in initial/early interphase (Gernand et al. 2005; Kim et al. 2002; Subrahma and Kasha 1973). Thus, consecutive cell divisions (mitosis) during the process of embryo development results in chromosomal elimination of male parent that give rise to a haploid embryo. Barclay (1975) successfully intercrossed hexaploid wheat with *H. bulbosum* to develop haploids in hexaploid wheat. However, genotypic dependence to some extent and lack of crossability (crossability barrier) with *H. bulbosum* are among the major hurdles to use Bulbosum method in wheat (Snape et al. 1979).

2.3.2   Haploids Using Maize as a Pollen Donor

The barriers with respect to crossability have not been reported between crosses of maize and wheat. Crossability genes that have been mapped on wheat chromosomes i.e., *Kr1* on 5BL, *Kr2* on 5AL, *Kr3* on 5D, and *Kr4* on 1A are not sensitive; thereby, do not create any barriers/hindrance. High frequency of green haploid plants has been obtained using maize as pollen donor (wide hybridizer) not only in wheat but also in barley (Furusho et al. 1991) and triticale (Wedzony et al. 1998). Genotypic dependence (to some extent), growth and developmental conditions (Campbell et al. 2001) and emasculation method (Knox et al. 2000) have been illustrated as major aspects affecting the frequency of green plants to a greater extent in wheat. The spray of growth regulators following pollination (2,4-D or/and Dicamba) or injection in last internode have significantly enhanced embryo production (Wedzony et al. 1998). The frequency of haploid production in oat using maize as a wide hybridizer is low (Rines and Dahleen 1990) because maize chromosome are not completely or entirely removed/eliminated during caryopsis. It often results in the production of more polyhaploids than haploids, though; these polyhaploids have been exploited in other genetic studies due to their use in the production of aneuploids (Rines 2003). Few other species have also been used as a wide hybridizer that includes *Zea mays* sp. Mexicana, pearl millet, Job's-tears (*Coix lachryma-jobi* L.), and sorghum (Inagaki and Mujeeb-Kazi 1997; Mochida and Tsujimoto 2001; Riera-Lizarazu et al. 1993; Ushiyama et al. 1991).

2.3.3   *Haploids Using* Solanum phureja *and Maize Inducer Lines*

In *Solanum tuberosum* (cultivated tetraploid potato), haploid plants are produced by crossing it with a diploid species, *S. phureja*, used as a pollen donor. The cross gives rise to a functional endosperm that result from the union of both sperm nuclei with central wall of ovule. Maine (2003) explained that this fusion initiates growth and development of unfertilized egg via parthenogenesis. The percentage of haploid embryos produced from a cross between *S. tuberosum* and *S. phureja* is extremely low, however, the haploid embryos can be simply differentiated from hybrid ones using a colored gene marker. The colored gene markers have been incorporated in the haploid embryo by male parent (pollen donor). A similar technique of color gene marker is also being used in maize to produce haploid plants that involves crossing with haploid inducer line that transmit colored (scorable) gene markers like lec1 promoter driving CRC, anthocyanin gene, R-nj, and GFP (US Patent 20060185033). The genotypes RWS (Geiger and Gordillo 2009) and stock 6 (Eder and Chalyk 2002) have been used commercially to produce haploids on a larger scale in maize.

References

Abadie JC, Puttsepp U, Gebauer G, Faccio A, Bonfante P, Selosse MA (2006) *Cephalanthera longifolia* (Neottieae, Orchidaceae) is mixotrophic: a comparative study between green and nonphotosynthetic individuals. Can J Bot 84(9):1462–1477. doi:10.1139/b06-101

Agache S, Bachelier B, Debuyser J, Henry Y, Snape J (1989) Genetic analysis of anther culture response in wheat using aneuploid, chromosome substitution and translocation lines. Theor Appl Genet 77(1):7–11. doi:10.1007/bf00292308

Agarwal PK, Bhojwani SS (1993) Enhanced pollen grain embryogenesis and plant-regeneration in anther cultures of *Brassica juncea* cv PR-45. Euphytica 70(3):191–196. doi:10.1007/bf00023759

Aionesei T, Touraev A, Heberle-Bors E (2005) Pathways to microspore embryogenesis. In: Palmer CE, Keller WA, Kasha KJ (eds) Biotechnology in agriculture and forestry, vol 56. Springer, Berlin, pp 11–34

Aldemita RR, Zapata FJ (1991) Anther culture of rice - effects of radiation and media components on callus induction and plant-regeneration. Cereal Res Commun 19(1–2):9–32

Almoguera C, Jordano J (1992) Developmental and environmental concurrent expression of sun-flower dry-seed-stored low-molecular-weight heat-shock protein and lea messenger-RNAs. Plant Mol Biol 19(5):781–792. doi:10.1007/bf00027074

Antoine MS, Beckert M (1997) Spontaneous versus colchicine-induced chromosome doubling in maize anther culture. Plant Cell Tissue Organ Cult 48(3):203–207. doi:10.1023/a:1005840400121

Arnison PG, Keller WA (1990) A survey of anther culture response of *B. oleracea* L. cultivars grown under field conditions. Plant Breed 104:125–133

Asif M, François E, Aakash G, Eric A, Harpinder R, Dean S (2013) Organelle antioxidants improve microspore embryogenesis in wheat and triticale. In Vitro Cell Dev Biol Plant. doi:10.1007/s11627-013-9514-z

Atak Q, Celik O, Olgun A, Alikamanoglu S, Rzakoulieva A (2007) Effect of magnetic field on peroxidase activities of soybean tissue culture. Biotechnol Biotechnol Equip 21(2):166–171

Bajaj YPS (1980) Enhancement of the in-vitro development of triticale embryos by the endosperm of durum wheat *Triticum durum*. Cereal Res Commun 8(2):359–364

Bajaj YPS (1990) In vitro production of haploids and their use in cell genetics and plant breeding. In: Bajaj YPS (ed) Haploids in crop improvement, vol 12, Biotechnology in agriculture and forestry. Springer, Berlin, pp 1–44

Bajaj YPS, Reinert J, Heberle E (1977) Factors enhancing in vitro production of haploid plants in anthers and isolated microspores. In: Gautheret RJ (ed) La culture des tissus et des cellules des vegetaux. Masson, Paris, pp 47–58

Ball ST, Zhou HP, Konzak CF (1993) Influence of 2,4-D, IAA, and duration of callus induction in anther cultures of spring wheat. Plant Sci 90(2):195–200. doi:10.1016/0168-9452(93)90240-z

Barany I, Testillano PS, Mityko J, Risueno MC (2001) The switch of the microspore developmental programme in *Capsicum* involves HSP70 expression and leads to the production of haploid plants. Int J Dev Biol 45:S39–S40

Barclay IR (1975) High-frequencies of haploid production in wheat (*Triticum aestivum*) by chromosome elimination. Nature 256(5516):410–411. doi:10.1038/256410a0

Barro F, Martin A (1999) Response of different genotypes of *Brassica carinata* to microspore culture. Plant Breed 118(1):79–81. doi:10.1046/j.1439-0523.1999.118001079.x

Barwale UB, Widholm JM (1987) Somaclonal variation in plants regenerated from cultures of soybean. Plant Cell Rep 6(5):365–368

Bernard S (1977) Study of some factors contributing to the success of androgenesis by in vitro anther culture in hexaploid triticale. Annales de l' Amelioration des Plantes 27:639–656

Binarova P, Hause G, Cenklova V, Cordewener JHG, Campagne MMV (1997) A short severe heat shock is required to induce embryogenesis in late bicellular pollen of Brassica napus L. Sex Plant Reprod 10(4):200–208. doi:10.1007/s004970050088

Bishnoi U, Jain RK, Rohilla JS, Chowdhury VK, Gupta KR, Chowdhury JB (2000) Anther culture of recalcitrant indica × Basmati rice hybrids. Euphytica 114(2):93–101. doi:10.1023/A:1003915331143

Blakeslee AF (1939) The present and potential service of chemistry to plant breeding. Am J Bot 26(3):163–172. doi:10.2307/2436533

Branch WD, Kvien CK (1992) Cytoplasmically inherited albinism in peanut seedlings. J Hered 83(6):455–457

Broughton S (2008) Ovary co-culture improves embryo and green plant production in anther culture of Australian spring wheat (*Triticum aestivum* L.). Plant Cell Tissue Organ Cult 95(2):185–195. doi:10.1007/s11240-008-9432-7

Campbell AW, Griffin WB, Burritt DJ, Conner AJ (2001) The importance of light intensity for pollen tube growth and embryo survival in wheat x maize crosses. Ann Bot 87(4):517–522. doi:10.1006/anbo.2000.1363

Caredda S, Clement C (1999) Androgenesis and albinism in *Poaceae*: influence of genotype and carbohydrates. In: Clement C, Pacini E, Audran JC (eds) Anther and pollen: from biology to biotechnology. Springer, Berlin, pp 211–226

Caredda S, Doncoeur C, Devaux P, Sangwan RS, Clement C (2000) Plastid differentiation during androgenesis in albino and non-albino producing cultivars of barley (*Hordeum vulgare* L.). Sex Plant Reprod 13(2):95–104. doi:10.1007/s004970000043

Caredda S, Devaux P, Sangwan RS, Proult I, Clement C (2004) Plastid ultrastructure and DNA related to albinism in androgenetic embryos of various barley (*Hordeum vulgare*) cultivars. Plant Cell Tissue Organ Cult 76(1):35–43. doi:10.1023/a:1025812621775

Casimiro I, Beeckman T, Graham N, Bhalerao R, Zhang HM, Casero P, Sandberg G, Bennett MJ (2003) Dissecting Arabidopsis lateral root development. Trends Plant Sci 8(4):165–171. doi:10.1016/s1360-1385(03)00051-7

Chanana NP, Dhawan V, Bhojwani SS (2005) Morphogenesis in isolated microspore cultures of *Brassica juncea*. Plant Cell Tissue Organ Cult 83(2):169–177. doi:10.1007/s11240-005-4855-x

Chaudhary HK, Dhaliwal I, Singh S, Sethi GS (2003) Genetics of androgenesis in winter and spring wheat genotypes. Euphytica 132(3):311–319. doi:10.1023/a:1025094606482

Chen CC (1977) Invitro development of plants from microspores of rice. In Vitro Cell Dev Biol Plant 13(8):484–489

Chen X-W, Cistue L, Munoz-Amatriain M, Sanz M, Romagosa I, Castillo A-M, Valles M-P (2007) Genetic markers for doubled haploid response in barley. Euphytica 158(3):287–294. doi:10.1007/s10681-006-9310-5

Chiancone B, Tassoni A, Bagni N, Germana MA (2006) Effect of polyamines on in vitro anther culture of *Citrus clementina* Hort. ex Tan. Plant Cell Tissue Organ Cult 87(2):145–153. doi:10.1007/s11240-006-9149-4

Cho MS, Zapata FJ (1990) Plant-regeneration from isolated microspore of indica rice. Plant Cell Physiol 31(6):881–885

Chory J, Aguilar N, Peto CA (1991) The phenotype of *Arabidopsis thaliana* DET1 mutants suggests a role for cytokinins in greening. In: Jenkins GI, Schuch W (eds) Molecular biology of plant development. Society for Experimental Biology, Cambridge, pp 21–29

Chu CC (1981) The N6 medium and its applications to anther culture of cereal crops. In: Proceedings of symposium on plant tissue culture, Beijing, pp 43–50

Chu CC, Hill RD, Brulebabel AL (1990) High-frequency of pollen embryoid formation and plant-regeneration in triticum-aestivum l on monosaccharide containing media. Plant Sci 66(2):255–262. doi:10.1016/0168-9452(90)90211-6

Cistué L, Vallés MP, Echavarri B, Sanz J, Castillo A (2003) Barley anther culture. In: Maluszynski M, Kasha KJ, Forster BP, Szarejko I (eds) Doubled haploid production in crop plants: a manual. Kluwer Academic, Dordrecht, pp 29–34

Cistue L, Romagosa I, Batlle F, Echavarri B (2009) Improvements in the production of doubled haploids in durum wheat (*Triticum turgidum* L.) through isolated microspore culture. Plant Cell Rep 28(5):727–735. doi:10.1007/s00299-009-0690-6

Clewer HWB, Green SJ, Tutin F (1915) The constituents of *Gloriosa superba*. J Chem Soc 107:835–846. doi:10.1039/ct9150700835

Collins JL (1927) A low temperature type of albinism in barley. J Hered 18(7):331–334

Croser JS, Lulsdorf MM, Davies PA, Clarke HJ, Bayliss KL, Mallikarjuna N, Siddique KHM (2006) Toward doubled haploid production in the fabaceae: progress, constraints, and opportunities. Crit Rev Plant Sci 25(2):139–157. doi:10.1080/07352680600563850

Croser JS, Lulsdorf MM, Grewal RK, Usher KM, Siddique KHM (2011) Isolated microspore culture of chickpea (*Cicer arietinum* L.): induction of androgenesis and cytological analysis of early haploid divisions. In Vitro Cell Dev Biol Plant 47(3):357–368. doi:10.1007/s11627-011-9346-7

Dahleen LS (1999) Donor-plant environment effects on regeneration from barley embryo-derived callus. Crop Sci 39(3):682–685

Datta SK, Potrykus I (1998) Direct pollen embryogenesis in cereals. Experientia 44:43. doi:10.1104/pp. 121.3.687

Datta SK, Wenzel G (1987) Isolated microspore derived plant formation via embryogenesis in *Tricum aestivum* L. Plant Sci 48(1):49–54. doi:10.1016/0168-9452(87)90069-0

Davies PA (2003) Barley isolated microspore culture (IMC) method. In: Maluszynski M, Kasha KJ, Forster BP, Szarejko I (eds) Doubled haploid production in crop plants: a manual. Kluwer Academic, Dordrecht, pp 49–92

Devaux P, Pickering R (2005) Haploids in the Improvement of Poaceae. In: Don Palmer CE, Keller W, Kasha K (eds) Haploids in crop improvement II, vol 56. Biotechnology in agriculture and forestry. Springer, Berlin, pp 215–242. doi:10.1007/3-540-26889-8_11

Dhawi F, Al-Khayri JM (2008) Magnetic fields induce changes in photosynthetic pigments content in datepalm (*Phoenix dactylifera* L.) seedlings. Open Agric J 2:121–125

Dhooghe E, Van Laere K, Eeckhaut T, Leus L, Van Huylenbroeck J (2011) Mitotic chromosome doubling of plant tissues in vitro. Plant Cell Tissue Organ Cult 104(3):359–373. doi:10.1007/s11240-010-9786-5

Dias JCD (2001) Effect of incubation temperature regimes and culture medium on broccoli microspore culture embryogenesis. Euphytica 119(3):389–394

Duncan EJ, Heberle E (1976) Effect of temperature shock on nuclear phenomena in microspores of *Nicotiana tabacum* and consequently on plantlet production. Protoplasma 90(1–2):173–177. doi:10.1007/bf01276486

Dunwell JM (1976) Comparative-study of environmental and developmental factors which influence embryo induction and growth in cultured anthers of *Nicotiana tabacum*. Environ Exp Bot 16(2–3):109–118. doi:10.1016/0098-8472(76)90002-2

Dunwell JM (2010) Haploids in flowering plants: origins and exploitation. Plant Biotechnol J 8(4):377–424. doi:10.1111/j.1467-7652.2009.00498.x

Dunwell JM, Thurling N (1985) Role of sucrose in microspore embryo production in *Brassica napus* ssp oleifera. J Exp Bot 36(170):1478–1491. doi:10.1093/jxb/36.9.1478

Dwivedi SL, Nigam SN, Pandey SK, Gibbons RW (1984) Inheritance of albinism in certain interspecific and intersubspecific crosses in groundnut (*Arachis hypogaea* L). Euphytica 33(3):705–708. doi:10.1007/bf00021898

Echavarri B, Soriano M, Cistue L, Valles MP, Castillo AM (2008) Zinc sulphate improved microspore embryogenesis in barley. Plant Cell Tissue Organ Cult 93(3):295–301. doi:10.1007/s11240-008-9376-y

Eder J, Chalyk S (2002) In vivo haploid induction in maize. Theor Appl Genet 104(4):703–708. doi:10.1007/s00122-001-0773-4

Ekiz H, Konzak CF (1991) Nuclear and cytoplasmic control of anther culture response in wheat. 1. Analyses of alloplasmic lines. Crop Sci 31(6):1421–1427

Eudes F, Chugh A (2009) An overview of Triticale doubled haploids. In: Touraev A, Forster BP, Jain SM (eds) Advances in haploid production in higher plants. Springer, Dordrecht, pp 87–96. doi:10.1007/978-1-4020-8854-4_6

Fan Z, Armstrong KC, Keller WA (1988) Development of microspores invivo and invitro in *Brassica napus* L. Protoplasma 147(2–3):191–199. doi:10.1007/bf01403347

Ferrie AMR (2003) Microspore culture of Brassica species. In: Maluszynski M, Kasha KJ, Forster BP, Szarejko I (eds) Doubled haploid production in crop plants: a manual. Kluwer Academic, Dordrecht, pp 205–215

Ferrie AMR, Epp DJ, Keller WA (1995a) Evaluation of *Brassica rapa* L genotypes for microspore culture response and identification of a highly embryogenic line. Plant Cell Rep 14(9):580–584

Ferrie AMR, Palmer CE, Keller WA (1995b) Haploid embryogenesis. In: Thorpe TA (ed) In vitro embryogenesis in plants. Kluwer Academic, Dordrecht, pp 309–344

Ferrie AMR, Taylor DC, MacKenzie SL, Keller WA (1999) Microspore embryogenesis of high sn-2 erucic acid *Brassica oleracea* germplasm. Plant Cell Tissue Organ Cult 57(2):79–84. doi:10.1023/a:1006325431653

Filonova LH, Bozhkov PV, Brukhin VB, Daniel G, Zhivotovsky B, von Arnold S (2000) Two waves of programmed cell death occur during formation and development of somatic embryos in the gymnosperm, Norway spruce. J Cell Sci 113(24):4399–4411

Fletcher R, Coventry J, Kott L (1998) Manual for microspore culture technique for *Brassica napus*. Technical Bulletin, Department of Plant Agriculture, University of Guelph, Guelph, ON

Francis D (2007) The plant cell cycle - 15 years on. New Phytol 174(2):261–278. doi:10.1111/j.1469-8137.2007.02038.x

Furusho M, Suenaga K, Nakajima K (1991) Production of haploid barley plants from barley x maize and barley x italian ryegrass crosses. Jpn J Breed 41(1):175–179

Gagliardi D, Breton C, Chaboud A, Vergne P, Dumas C (1995) Expression of heat shock factor and heat shock protein 70 genes during maize pollen development. Plant Mol Biol 29(4):841–856. doi:10.1007/bf00041173

Gaillard A, Vergne P, Beckert M (1991) Optimization of maize microspore isolation and culture conditions for reliable plant regeneration. Plant Cell Rep 10(2):55–58. doi:10.1007/BF00236456

Garrido D, Vicente O, Heberlebors E, Rodriguezgarcia MI (1995) Cellular-changes during the acquisition of embryogenic potential in isolated pollen grains of *Nicotiana tabacum*. Protoplasma 186(3–4):220–230. doi:10.1007/bf01281332

Geiger HH, Gordillo GA (2009) Doubled haploids in hybrid maize breeding. Maydica 54(4):485–499

Gemes-Juhasz A, Balogh P, Ferenczy A, Kristof Z (2002) Effect of optimal stage of female game-tophyte and heat treatment on in vitro gynogenesis induction in cucumber (*Cucumis sativus* L.). Plant Cell Rep 21(2):105–111. doi:10.1007/s00299-002-0482-8

Germana MA (2007) Haploidy. Citrus Genet Breed Biotechnol. doi:10.1079/9780851990194.0167

Germana MA (2011a) Anther culture for haploid and doubled haploid production. Plant Cell Tissue Organ Cult 104(3):283–300. doi:10.1007/s11240-010-9852-z

Germana MA (2011b) Gametic embryogenesis and haploid technology as valuable support to plant breeding. Plant Cell Rep 30(5):839–857. doi:10.1007/s00299-011-1061-7

Germana MA, Chiancone B (2003) Improvement of *Citrus clementina* Hort. ex Tan. microspore-derived embryoid induction and regeneration. Plant Cell Rep 22(3):181–187. doi:10.1007/s00299-003-0669-7

Germana MA, Crescimanno FG, Motisi A (2000) Factors affecting androgenesis in *Citrus clementina* Hort. ex Tan. Adv Hortic Sci 14(2):43–51

Gernand D, Rutten T, Varshney A, Rubtsova M, Prodanovic S, Bruss C, Kumlehn J, Matzk F, Houben A (2005) Uniparental chromosome elimination at mitosis and interphase in wheat and pearl millet crosses involves micronucleus formation, progressive heterochromatinization, and DNA fragmentation. Plant Cell 17(9):2431–2438. doi:10.1105/tpc.105.034249

Gland A, Lichter R, Schweiger HG (1988) Genetic and exogenous factors affecting embryogenesis in isolated microspore cultures of *Brassica napus* L. J Plant Physiol 132(5):613–617

Greplova M, Polzerova H, Domkarova J (2009) Intra- and inter-specific crosses of Solanum materials after mitotic polyploidization in vitro. Plant Breed 128(6):651–657. doi:10.1111/j.1439-0523.2009.01632.x

Gu HH, Zhou WJ, Hagberg P (2003a) High frequency spontaneous production of doubled haploid plants in microspore cultures of *Brassica rapa* ssp chinensis. Euphytica 134(3):239–245. doi:10.1023/B:EUPH.0000004945.01455.6d

Gu HH, Zhu J, Zhang G, Zhou W (2003b) Microspore culture and ploidy identification of regener-ated plant in Chinese flowering cabbage (*Brassica rapa* ssp. parachinensis). J Agric Biol 11(6):572–576

Guha S, Maheshwari SC (1964) In vitro production of embryos from anthers of datura. Nature 204(495):497. doi:10.1038/204497a0

Guo YD, Pulli S (2000a) An efficient androgenic embryogenesis and plant regeneration method through isolated microspore culture in timothy (*Phleum pratense* L.). Plant Cell Rep 19(8):761–767

Guo YD, Pulli S (2000b) Isolated microspore culture and plant regeneration in rye (*Secale cereale* L.). Plant Cell Rep 19(9):875–880

Gustafson VD, Baenziger PS, Wright MS, Stroup WW, Yen Y (1995) Isolated wheat microspore culture. Plant Cell Tissue Organ Cult 42(2):207–213. doi:10.1007/bf00034239

Hansen M (2000) ABA treatment and desiccation of microspore-derived embryos of cabbage (*Brassica oleracea* ssp capitata L.) improves plant development. J Plant Physiol 156(2):164–167

Hansen M (2003) Protocol for microspore culture in Brassica. In: Maluszymski M, Kasha KJ, Forster BP, Szarejko I (eds) Doubled haploid production in crop plants: a manual. Kluwer Academic, Dordrecht, pp 217–222

Harada H, Kyo M, Imamura J (1988) The induction of embryogenesis in nicotiana immature pollen in culture. In: Bock G, Marsh J (eds) Applications of plant cell and tissue culture. Ciba Foundation Symposium, Kyoto, Japan, pp 59–69

He P, Shen LH, Lu CF, Chen Y, Zhu LH (1998) Analysis of quantitative trait loci which contribute to anther culturability in rice (*Oryza sativa* L.). Mol Breed 4(2):165–172. doi:10.1023/a:1009692221152

Heberle-Bors E (1984) Pollen embryogenesis a model system for the life cycle of higher plants. J Embryol Exp Morphol 82(suppl):17

Heberle-Bors E (1989) Isolated pollen culture in tobacco: plant reproductive development in a nutshell. Sex Plant Reprod 2(1):1–10

Ho KM, Kasha KJ (1975) Genetic-control of chromosome elimination during haploid formation in barley. Genetics 81(2):263–275

Hoekstra S, Vanzijderveld MH, Louwerse JD, Heidekamp F, Vandermark F (1992) Anther and microspore culture of *Hordeum vulgare* L cv Igri. Plant Sci 86(1):89–96. doi:10.1016/0168-9452(92)90182-1

Hu T, Kasha KJ (1997) Improvement of isolated microspore culture of wheat (*Triticum aestivum* L) through ovary co-culture. Plant Cell Rep 16(8):520–525

Hu TC, Kasha KJ (1999) A cytological study of pretreatments used to improve isolated microspore cultures of wheat (*Triticum aestivum* L.) cv. Chris. Genome 42(3):432–441. doi:10.1139/gen-42-3-432

Huang B, Bird S, Kemble R, Simmonds D, Keller W, Miki B (1990) Effects of culture density, conditioned medium and feeder cultures on microspore embryogenesis in *Brassica napus* L cv Topas. Plant Cell Rep 8(10):594–597

Hunter CP (1987) Plant regeneration method. European Patent

Immonen S, Anttila H (2000) Media composition and anther plating for production of androgenetic green plants from cultivated rye (*Secale cereale* L.). J Plant Physiol 156(2):204–210

Inagaki MN, Mujeeb-Kazi A (1997) Production of polyhaploids of hexaploid wheat using stored pearl millet pollen. Wheat: prospects for global improvement. In: Proceedings of the 5th international wheat conference, Ankara, Turkey, 10–14 June 1996

Indrianto A, Heberle-Bors E, Touraev A (1999) Assessment of various stresses and carbohydrates for their effect on the induction of embryogenesis in isolated wheat microspores. Plant Sci 143(1):71–79. doi:10.1016/s0168-9452(99)00022-9

Indrianto A, Barinova I, Touraev A, Heberle-Bors E (2001) Tracking individual wheat microspores in vitro: identification of embryogenic microspores and body axis formation in the embryo. Planta 212(2):163–174. doi:10.1007/s004250000375

Iqbal MCM, Wijesekara KB (2007) A brief temperature pulse enhances the competency of microspores for androgenesis in *Datura metel*. Plant Cell Tissue Organ Cult 89(2–3):141–149. doi:10.1007/s11240-007-9222-7

Jacquard C, Asakaviciute R, Hamalian AM, Sangwan RS, Devaux P, Clement C (2006) Barley anther culture: effects of annual cycle and spike position on microspore embryogenesis and albinism. Plant Cell Rep 25(5):375–381. doi:10.1007/s00299-005-0070-9

Jacquard C, Nolin F, Hecart C, Grauda D, Rashal I, Dhondt-Cordelier S, Sangwan RS, Devaux P, Mazeyrat-Gourbeyre F, Clement C (2009) Microspore embryogenesis and programmed cell death in barley: effects of copper on albinism in recalcitrant cultivars. Plant Cell Rep 28(9):1329–1339. doi:10.1007/s00299-009-0733-z

Jensen CJ (1974) Chromosome doubling techniques in haploids. In: Kasha KJ (ed) Haploids in higher plants, advances and potential III. Methods of chromosome doubling. University of Guelph, Guelph, ON, pp 153–190

Joersbo M, Jorgensen RB, Olesen P (1990) Transient electropermeabilization of barley (*Hordeum vulgare* L) microspores to propidium iodide. Plant Cell Tissue Organ Cult 23(2):125–129. doi:10.1007/bf00035832

Kao KN (1981) Plant formation from barley anther cultures with ficoll media. Zeitschrift Fur Pflanzenphysiologie 103(5):437–443

Kao KN, Saleem M, Abrams S, Pedras M, Horn D, Mallard C (1991) Culture conditions for induction of green plants from barley microspores by anther culture methods. Plant Cell Rep 9(11):595–601

Karsai I, Bedo Z (1997) Effect of carbohydrate content on the embryoid and plant production in triticale anther culture. Cereal Res Commun 25(2):109–116

Kasha KJ, Kao KN (1970) High frequency haploid production in barley (*Hordeum vulgare* L.). Nature 225(5235):874–876. doi:10.1038/225874a0

Kasha KJ, Maluszynski M (2003) Production of doubled haploids in crop plants. In: Maluszymski M, Kasha KJ, Forster BP, Szarejko I (eds) Doubled haploid production in crop plants: a manual. Kluwer Academic, Dordrecht, pp 1–4

Kasha KJ, Simion E, Oro R, Yao QA, Hu TC, Carlson AR (2001) An improved in vitro technique forisolatedmicrosporecultureofbarley.Euphytica120(3):379–385.doi:10.1023/a:1017564100823

Kasha KJ, Simion E, Miner M, Letarte J, Hu TC (2003a) Haploid wheat isolated microspore culture protocol. In: Maluszymski M, Kasha KJ, Forster BP, Szarejko I (eds) Doubled haploid production in crop plants: a manual. Kluwer Academic, Dordrecht, pp 77–81

Kasha KJ, Simion E, Oro R, Shim YS (2003b) Barley isolated microspore culture protocol. In: Maluszymski M, Kasha KJ, Forster BP, Szarejko I (eds) Doubled haploid production in crop plants: a manual. Kluwer Academic, Dordrecht, pp 43–48

Keller W, Arnison PG, Cardy BJ (1987a) Haploids from gametophytic cells recent developments and future prospects. In: Green CE et al (eds) Plant biology. Liss, New York, pp 223–242

Keller W, Fan Z, Pechan P, Long N, Grainger J (1987b) An efficient method for culture of isolated microspores in *Brassica napus* (Poster). In: 7th International rapeseed congress, Poznan, 11–14 May 1987, p 71

Kieffer M, Fuller MP, Chauvin JE, Schlesser A (1993) Anther culture of kale (*Brassica oleracea* L convar acephala (dc) alef). Plant Cell Tissue Organ Cult 33(3):303–313. doi:10.1007/bf02319016

Kim NS, Armstrong KC, Fedak G, Ho K, Park NI (2002) A microsatellite sequence from the rice blast fungus (*Magnaporthe grisea*) distinguishes between the centromeres of *Hordeum vulgare* and *H. bulbosum* in hybrid plants. Genome 45(1):165–174. doi:10.1139/g01-129

Kim M, Jang I-C, Kim J-A, Park E-J, Yoon M, Lee Y (2008) Embryogenesis and plant regeneration of hot pepper (*Capsicum annuum* L.) through isolated microspore culture. Plant Cell Rep 27(3):425–434. doi:10.1007/s00299-007-0442-4

Kiviharju E, Laurila J, Lehtonen M, Tanhuanpaa P, Manninen O (2004) Anther culture properties of oat x wild red oat progenies and a search for RAPD markers associated with anther culture ability. Agric Food Sci 13(1–2):151–162. doi:10.2137/1239099041838094

Knox RE, Clarke JM, DePauw RM (2000) Dicamba and growth condition effects on doubled haploid production in durum wheat crossed with maize. Plant Breed 119(4):289–298. doi:10.1046/j.1439-0523.2000.00498.x

Konzak CF, Polle EA, Liu W, Zheng Y (1999) Methods for generating doubled-haploid plants. US Patent US6362393. PCT/US99/19498

Kumar HGA, Ravishankar BV, Murthy HN (2004) The influence of polyamines on androgenesis of *Cucumis sativus* L. Eur J Hortic Sci 69(5):201–205

Kyo M, Harada H (1986) Control of the developmental pathway of tobacco pollen invitro. Planta 168(4):427–432. doi:10.1007/bf00392260

Kyo M, Harada H (1990) Specific phosphoproteins in the initial period of tobacco pollen embryogenesis. Planta 182(1):58–63

Lantos C, Paricsi S, Zofajova A, Weyen J, Pauk J (2006) Isolated microspore culture of wheat (*Triticum aestivum* L.) with Hungarian cultivars. Acta Biologica Szegediensis 50(1–2):31–35

Lantos C, Juhasz AG, Somogyi G, Otvos K, Vagi P, Mihaly R, Kristof Z, Somogyi N, Pauk J (2009) Improvement of isolated microspore culture of pepper (*Capsicum annuum* L.) via co-culture with ovary tissues of pepper or wheat. Plant Cell Tissue Organ Cult 97(3):285–293. doi:10.1007/s11240-009-9527-9

Larsen ET, Tuvesson IKD, Andersen SB (1991) Nuclear genes affecting percentage of green plants in barley (*Hordeum vulgare* L) anther culture. Theor Appl Genet 82(4):417–420. doi:10.1007/bf00588593

Letarte J, Simion E, Miner M, Kasha KJ (2006) Arabinogalactans and arabinogalactan-proteins induce embryogenesis in wheat (*Triticum aestivum* L.) microspore culture. Plant Cell Rep 24(12):691–698. doi:10.1007/s00299-005-0013-5

Li HC, Devaux P (2003) High frequency regeneration of barley doubled haploid plants from isolated microspore culture. Plant Sci 164(3):379–386. doi:10.1016/s0168-9452(02)00424-7

Liang GH, Xu A, Hoangtang (1987) Direct generation of wheat haploids via anther culture. Crop Sci 27(2):336–339

Lichter R (1982) Induction of haploid plants from isolated pollen of *Brassica napus*. Zeitschrift Fur Pflanzenphysiologie 105(5):427–434

Lindstorm EW (1929) A haploid mutant in tomato. J Hered 17:351–357

Low D, Brandle K, Nover L, Forreiter C (2000) Cytosolic heat-stress proteins Hsp17.7 class I and Hsp17.3 class II of tomato act as molecular chaperones in vivo. Planta 211(4):575–582. doi:10.1007/s004250000315

Lu R, Wang Y, Sun Y, Shan L, Chen P, Huang J (2008) Improvement of isolated microspore culture of barley (*Hordeum vulgare* L.): the effect of floret co-culture. Plant Cell Tissue Organ Cult 93(1):21–27. doi:10.1007/s11240-008-9338-4

Luckett DJ (1989) Colchicine mutagenesis is associated with substantial heritable variation in cotton. Euphytica 42(1–2):177–182. doi:10.1007/bf00042630

Luk YSF, Ignatova SA, Sozinov AA (1983) Use of in vitro techniques for producing haploids in barley and triticale. Tagungsbericht, Akademie der Landwirtschaftswissenschaften der Deutschen Demokratischen Republik (207):41–48

Maine D (2003) Potato haploid technologies. In: Maluszynski M, Kasha KJ, Forster BP, Szarejko I (eds) Doubled haploid production in crop plants: a manual. Kluwer Academic, Dordrecht, pp 241–247

Marciniak K, Kaczmarek Z, Adamski T, Surma M (2003) The anther-culture response of triticale line X tester progenies. Cell Mol Biol Lett 8(2):343–351

Mejza SJ, Morgant V, Dibona DE, Wong JR (1993) Plant-regeneration from isolated microspores of *Triticum aestivum*. Plant Cell Rep 12(3):149–153

Miah MAA, Earle ED, Khush GS (1985) Inheritance of callus formation ability in anther cultures of rice, *Oryza sativa* L. Theor Appl Genet 70(2):113–116

Mochida K, Tsujimoto H (2001) Production of wheat doubled haploids by pollination with Job's tears (*Coix lachryma*-jobi L.). J Hered 92(1):81–83. doi:10.1093/jhered/92.1.81

Moieni A, Sarrafi A (1995) Genetic-analysis for haploid-regeneration responses of hexaploid-wheat anther cultures. Plant Breed 114(3):247–249. doi:10.1111/j.1439-0523.1995.tb00803.x

Morejohn LC, Fosket DE (1984) Inhibition of plant microtubule polymerization invitro by the phosphoric amide herbicide *Amiprophos methyl*. Science 224(4651):874–876. doi:10.1126/science.224.4651.874

Mouritzen P, Holm PB (1994) Chloroplast genome breakdown in microspore cultures of barley (*Hordeum vulgare* L) occurs primarily during regeneration. J Plant Physiol 144(4–5):586–593

Murashige T, Skoog F (1962) A revised medium for rapid growth and bio assays with tobacco tissue cultures. Physiol Plant 15(3):473–497. doi:10.1111/j.1399-3054.1962.tb08052.x

Neuffer MG, Coe EH, Wessler SR (1997) Mutants of maize. Cold Spring Harbor Laboratory Press, Plainville, NY

Nimura M, Kato J, Horaguchi H, Mii M, Sakai K, Katoh T (2006) Induction of fertile amphidiploids by artificial chromosome-doubling in interspecific hybrid between *Dianthus caryophyllus* L. and *D. japonicus* thunb. Breed Sci 56(3):303–310. doi:10.1270/jsbbs.56.303

Nishiyam I, Motoyosh F (1966) Cytogenetic studies in avena 16. Chlorophyll formation in albino sand oats under certain culture conditions. Jpn J Genet 41(5):403. doi:10.1266/jjg.41.403

Nishiyama I, Motoyoshi F (1966) Cytogenetic studies in Avena. XVI. Chlorophyll formation in albino sand oats under certain culture conditions. Jpn J Genet 41:403–411

Nitsch C (1974) Pollen culture - a new technique for mass production of haploid and homozygous plants. In: Kasha KJ (ed) Haploids in higher plants, advances and potential. II. Methods of producing haploids. University of Guelph, Guelph, ON, pp 123–135

Niu YZ, Liu YZ, Wang LZ, Yuang YX, Li SC, Fang QJ (1999) A preliminary study on isolated microspore culture and plant regeneration of resynthesized *Brassica napus*. J Sichuan Agric Univ 17:167–171

Obert B, Barnabas B (2004) Colchicine induced embryogenesis in maize. Plant Cell Tissue Organ Cult 77(3):283–285. doi:10.1023/b:ticu.0000018399.60106.33

Ohkawa Y, Belvis E, Keller WA (1987) Validity study and microspore culture method in *Brassica napus*. Cruciferae Newslett 13:75

Olsen FL (1987) Induction of microspore embryogenesis in cultured anthers of *Hordeum vulgare* - the effects of ammonium-nitrate, glutamine and asparagine as nitrogen-sources. Carlsberg Res Commun 52(6):393–404. doi:10.1007/bf02907527

Olsen FL (1991) Isolation and cultivation of embryogenic microspores from barley (*Hordeum vulgare* L). Hereditas 115(3):255–266

Omran SA, Guerra-Sanz JM, Garrido Cardenas JA (2008) Methodology of tetraploid induction and expression of microsatellite alleles in triploid watermelon. In: Pitrat M (ed) Cucurbitaceae 2008: Proceedings of the IXth Eucarpia meeting on genetics and breeding of cucurbitaceae, 21–24 May 2008. INRA, Avignon, France, pp 381–384

Osolnik B, Bohanec B, Jelaska S (1993) Stimulation of androgenesis in white cabbage (*Brassica oleracea* var capitata) anthers by low-temperature and anther dissection. Plant Cell Tissue Organ Cult 32(2):241–246. doi:10.1007/bf00029849

Otani M, Shimada T (1994) Pollen embryo formation and plant regeneration from cultured anthers of tetraploid wheat. J Genet Breed 48(1):103–106

Pan JL, Gao GH, Ban H (1983) Initial patterns of androgenesis in wheat anther culture. Acta Bot Sin 25:34–39

Parcellier A, Gurbuxani S, Schmitt E, Solary E, Garrido C (2003) Heat shock proteins, cellular chaperones that modulate mitochondrial cell death pathways. Biochem Biophys Res Commun 304(3):505–512. doi:10.1016/s0006-291x(03)00623-5

Pauk J, Puolimatka M, Toth KL, Monostori T (2000) In vitro androgenesis of triticale in isolated microspore culture. Plant Cell Tissue Organ Cult 61(3):221–229. doi:10.1023/a:1006416116366

Pauk J, Mihaly R, Monostori T, Puolimatka M (2003) Protocol of triticale (x *Triticosecale Wittmack*) microspore culture. In: Maluszynski M, Kasha KJ, Forster BP, Szarejko I (eds) Doubled haploid production in crop plants: a manual. Kluwer Academic, Dordrecht, pp 129–134

Pechan PM, Keller WA (1989) Induction of microspore embryogenesis in *Brassica napus* L by gamma-irradiation and ethanol stress. In Vitro Cell Dev Biol Plant 25(11):1073–1074

Pechan PM, Smykal P (2001) Androgenesis: affecting the fate of the male gametophyte. Physiol Plant 111(1):1–8. doi:10.1034/j.1399-3054.2001.1110101.x

Pei XS (1985) Polyploidy induction and breeding. Shanghai Science and Technology Press, Shanghai, pp 95–98

Petolino JF, Jones AM, Thompson SA (1988) Selection for increased anther culture response in maize. Theor Appl Genet 76(1):157–159. doi:10.1007/bf00288847

Phippen C, Ockendon DJ (1990) Genotype, plant, bud size and media factors affecting anther culture of cauliflowers (*Brassica oleracea* var botrytis). Theor Appl Genet 79(1):33–38

Pickering RA (1984) The influence of genotype and environment on chromosome elimination in crosses between *Hordeum vulgare*-L x *Hordeum bulbosum* L. Plant Sci Lett 34(1–2):153–164. doi:10.1016/0304-4211(84)90138-x

Pingping Z, Ruochun Y, Zhiyou C, Zengliang Y (2011) Genotoxic effects of superconducting static magnetic fields (SMFs) on wheat (*Triticum aestivum*) pollen mother cells (PMCs). Plasma Sci Technol 9(2):241–247

Prem D, Gupta K, Sarkar G, Agnihotri A (2008) Activated charcoal induced high frequency micro-spore embryogenesis and efficient doubled haploid production in *Brassica juncea*. Plant Cell Tissue Organ Cult 93(3):269–282. doi:10.1007/s11240-008-9373-1

Puddephat IJ, Robinson HT, Smith BM, Lynn J (1999) Influence of stock plant pretreatment on gynogenic embryo induction from flower buds of onion. Plant Cell Tissue Organ Cult 57(2):145–148. doi:10.1023/a:1006312614874

Pulli S, Guo JD (2003) Microspore culture of rye. In: Maluszymski M, Kasha KJ, Forster BP, Szarejko I (eds) Doubled haploid production in crop plants: a manual. Kluwer Academic, Dordrecht, pp 151–154

Qu RD, Chen Y (1984) Pathways of androgenesis and observations on cultured pollen grains in rice *Oryza sativa* ssp keng. Acta Bot Sin 26(6):580–587

Raghavan V (1976) Role of generative cell in androgenesis in henbane. Science 191(4225):388–389. doi:10.1126/science.191.4225.388

Raghavan V (1978) Origin and development of pollen embryoids and pollen calluses in cultured anther segments of *Hyoscyamus niger* (henbane). Am J Bot 65(9):984–1002. doi:10.2307/2442686

Raghavan V (1986) Embryogenesis in angiosperms. A developmental and experimental study. Developmental and cell biology, vol 17. Cambridge University Press, Melbourne

Raghavan V (1990) From microspore to embryoid - faces of the angiosperm pollen grain, vol 9. Progress in plant cellular and molecular biology. Kluwer Academic, Dordrecht

Raina SK, Irfan ST (1998) High-frequency embryogenesis and plantlet regeneration from isolated microspores of indica rice. Plant Cell Rep 17(12):957–962

Rajyalakshmi K, Chowdhry CN, Maheshwari N, Maheshwari SC (1995) Anther culture response in some Indian wheat cultivars and the role of polyamines in induction of haploids. Phytomorphology 45(1–2):139–145

Ramirez C, Testillano PS, Castillo AM, Valles MP, Coronado MJ, Cistue L, Risueno MD (2001) The early microspore embryogenesis pathway in barley is accompanied by concrete ultrastruc-tural and expression changes. Int J Dev Biol 45:S57–S58

Randolph LF (1932) Some effects of high temperature on polyploidy and other variations in maize. Proc Natl Acad Sci USA 18:222–229. doi:10.1073/pnas.18.3.222

Redha A, Talaat A (2008) Improvement of green plant regeneration by manipulation of anther culture induction medium of hexaploid wheat. Plant Cell Tissue Organ Cult 92(2):141–146. doi:10.1007/s11240-007-9315-3

Reinbothe C, Buhr F, Pollmann S, Reinbothe S (2003a) In vitro reconstitution of light-harvesting POR-protochlorophyllide complex with protochlorophyllides a and b. J Biol Chem 278(2):807–815. doi:10.1074/jbc.M209738200

Reinbothe S, Pollmann S, Reinbothe C (2003b) In situ conversion of protochlorophyllide b to protochlorophyllide a in barley - evidence for a novel role of 7-formyl reductase in the prola-mellar body of etioplasts. J Biol Chem 278(2):800–806. doi:10.1074/jbc.M209737200

Reynolds TL (1993) A cytological analysis of microspores of *Triticum aestivum* (poaceae) during normal ontogeny and induced embryogenic development. Am J Bot 80(5):569–576. doi:10.2307/2445374

Riera-Lizarazu O, Mujeeb-Kazi A, William MDHM (1993) Maize (*Zea mays* L.) mediated polyhap-loid production in some Triticeae using a detached tiller method. J Genet Breed 46(4):335–346

Rines HW (2003) Oat haploids from wide hybridization. In: Maluszymski M, Kasha KJ, Forster BP, Szarejko I (eds) Doubled haploid production in crop plants: a manual. Kluwer Academic, Dordrecht, pp 155–160

Rines HW, Dahleen LS (1990) Haploid oat plants produced by application of maize pollen to emasculated oat florets. Crop Sci 30(5):1073–1078

Sabehat A, Lurie S, Weiss D (1998) Expression of small heat-shock proteins at low temperatures - a possible role in protecting against chilling injuries. Plant Physiol 117(2):651–658. doi:10.1104/pp. 117.2.651

Saidi N, Cherkaoui S, Chlyah A, Chlyah H (1997) Embryo formation and regeneration in *Triticum turgidum* ssp. durum anther culture. Plant Cell Tissue Organ Cult 51(1):27–33. doi:10.1023/a:1005765529154

Sangwan RS, Sangwannorreel BS (1987a) Biochemical cytology of pollen embryogenesis. Int Rev Cytol 107:221–272. doi:10.1016/s0074-7696(08)61077-3

Sangwan RS, Sangwannorreel BS (1987b) Ultrastructural cytology of plastids in pollen grains of certain androgenic and nonandrogenic plants. Protoplasma 138(1):11–22. doi:10.1007/bf01281180

Sari N, Abak K, Pitrat M (1999) Comparison of ploidy level screening methods in watermelon: *Citrullus lanatus* (Thunb.) Matsum. and Nakai. Sci Hortic 82(3–4):265–277. doi:10.1016/s0304-4238(99)00077-1

Sato S, Katoh N, Iwai S, Hagimori M (2002) Effect of low temperature pretreatment of buds or inflorescence on isolated microspore culture in *Brassica rapa* (syn. *B. campestris*). Breed Sci 52:23–26

Scott P, Lyne RL, Aprees T (1995) Metabolism of maltose and sucrose by microspores isolated from barley (*Hordeum vulgare* L). Planta 197(3):435–441

Segui-Simarro JM, Testillano PS, Risueno MC (2003) Hsp70 and Hsp90 change their expression and subcellular localization after microspore embryogenesis induction in *Brassica napus* L. J Struct Biol 142(3):379–391. doi:10.1016/s1047-8477(03)00067-4

Shariatpanahi ME, Bal U, Heberle-Bors E, Touraev A (2006) Stresses applied for the re-programming of plant microspores towards in vitro embryogenesis. Physiol Plant 127(4):519–534. doi:10.1111/j.1399-3054.2006.00675.x

Sibi ML, Kobaissi A, Shekafandeh A (2001) Green haploid plants from unpollinated ovary culture in tetraploid wheat(*Triticum durum* Defs.).Euphytica 122(2):351–359.doi:10.1023/a:1012991325228

Snape JW, Chapman V, Moss J, Blanchard CE, Miller TE (1979) Crossabilities of wheat-varieties with *Hordeum bulbosum*. Heredity 42:291–298. doi:10.1038/hdy.1979.32

Sopory SK, Munshi M (1997) Anther culture. In: Jain SM, Sopory SK, Veilleux RE (eds) In vitro haploid production in higher plants. Kluwer Academic, Dordrecht, pp 145–176

Soriano M, Cistue L, Castillo AM (2008) Enhanced induction of microspore embryogenesis after n-butanol treatment in wheat (*Triticum aestivum* L.) anther culture. Plant Cell Rep 27(5):805–811. doi:10.1007/s00299-007-0500-y

Subrahma NC, Kasha KJ (1973) Selective chromosomal elimination during haploid formation in barley following interspecific hybridization. Chromosoma 42(2):111–125. doi:10.1007/bf00320934

Sun CS (1978) Androgenesis of cereal crops. In: Proceedings of symposium on plant tissue culture. Science Press, Peking, pp 117–123

Sunderla N, Wicks FM (1971) Embryoid formation in pollen grains of *Nicotiana tabacum*. J Exp Bot 22(70):213–226. doi:10.1093/jxb/22.1.213

Sunderland N (1974) Anther culture as a means of haploid induction. In: Kasha KJ (ed) Haploids in higher plants, advances and potential II. Methods of producing haploids. University of Guelph, Guelph, ON, pp 91–122

Sunderland N, Dunwell JM (1977) Anther and pollen culture. In: Street HE (ed) Plant tissue and cell culture. Blackwell, Oxford, pp 223–265

Sunderland N, Roberts M (1977) New approach to pollen culture. Nature 270(5634):236–238. doi:10.1038/270236a0

Sunderland N, Roberts M (1979) Cold-pretreatment of excised flower buds in float culture of tobacco anthers. Ann Bot 43(4):405–414

Sunderland N, Xu ZH (1982) Shed pollen culture in *Hordeum vulgare*. J Exp Bot 33(136):1086–1095. doi:10.1093/jxb/33.5.1086

Sunderland N, Huang B, Hills GJ (1984) Disposition of pollen insitu and its relevance to anther pollen culture. J Exp Bot 35(153):521–530. doi:10.1093/jxb/35.4.521

Swanson EB, Coumans MP, Wu SC, Barsby TL, Beversdorf WD (1987) Efficient isolation of microspores and the production of microspore-derived embryos from *Brassica napus*. Plant Cell Rep 6(2):94–97

Takahata Y, Brown DCW, Keller WA (1991) Effect of donor plant-age and inflorescence age on microspore culture of *Brassica napus* L. Euphytica 58(1):51–55. doi:10.1007/bf00035339

Takamura T, Lim KB, Van Tuyl JM (2002) Effect of a new compound on the mitotic polyploidization of *Lilium longiflorum* and oriental hybrid lilies. In: VanHuylenbroeck J, VanBockstaele E, Debergh P (eds) Proceedings of the twentieth international eucarpia symposium - Section ornamentals: strategies for new ornamentals II, vol 572, Acta Hortic., pp 37–42

Telmer CA, Newcomb W, Simmonds DH (1993) Microspore development in *Brassica napus* and the effect of high-temperature on division invivo and invitro. Protoplasma 172(2–4):154–165. doi:10.1007/bf01379373

Testillano P, Georgiev S, Mogensen HL, Coronado MJ, Dumas C, Risueno MC, Matthys-Rochon E (2004) Spontaneous chromosome doubling results from nuclear fusion during in vitro maize induced microspore embryogenesis. Chromosoma 112(7):342–349. doi:10.1007/s00412-004-0279-3

Torp AM, Hansen AL, Andersen SB (2001) Chromosomal regions associated with green plant regeneration in wheat (*Triticum aestivum* L.) anther culture. Euphytica 119(3):377–387. doi:1 0.1023/a:1017554129904

Touraev A, Ilham A, Vicente O, HeberleBors E (1996a) Stress-induced microspore embryogenesis in tobacco: an optimized system for molecular studies. Plant Cell Rep 15(8):561–565

Touraev A, Indrianto A, Wratschko I, Vicente O, HeberleBors E (1996b) Efficient microspore embryogenesis in wheat (*Triticum aestivum* L) induced by starvation at high temperature. Sex Plant Reprod 9(4):209–215. doi:10.1007/bf02173100

Touraev A, Pfosser M, Vicente O, HeberleBors E (1996c) Stress as the major signal controlling the developmental fate of tobacco microspores: towards a unified model of induction of microspore/pollen embryogenesis. Planta 200(1):144–152

Touraev A, Pfosser M, Heberle-Bors E (2001) The microspore: a haploid multipurpose cell. Adv Bot Res 35:53–109. doi:10.1016/s0065-2296(01)35004-8

Touraev A, Forster BP, Jain SM (2009) Advances in haploid production in higher plants. Springer, Dordrecht

Tsay HS (1982) The microspore development and haploid embryogenesis of anther culture with five nitrogen doses to the donor tobacco plants. J Agric Res China 31(1):1–13

Tuvesson IKD, Pedersen S, Andersen SB (1989) Nuclear genes affecting albinism in wheat (*Triticum aestivum* L) anther culture. Theor Appl Genet 78(6):879–883

Ushiyama T, Shimizu T, Kuwabara T (1991) High-frequency of haploid production of wheat through intergeneric cross with teosinte. Jpn J Breed 41(2):353–357

Vagera J, Havranek P (1983) Regulation of androgenesis in *Nicotiana tabacum* L cv white burley and *Datura innoxia* mill effect of bivalent and trivalent iron and chelating substances. Biol Plant 25(1):5–14. doi:10.1007/bf02878260

Vagera J, Novotny J, Ohnoutkova L (2004) Induced androgenesis in vitro in mutated populations of barley, *Hordeum vulgare*. Plant Cell Tissue Organ Cult 77(1):55–61. doi:10.1023/B:TICU.0000016504.82810.a4

Van LK (2008) Interspecific hybridisation in woody ornamentals. Ghent University, Ghent

Vasil IK (1980) Androgenic haploids. Int Rev Cytol Suppl 11A:195–223

Wang CC, Chu CC, Sun CS, Wu SH, Yin KC, Hsu C (1973) The androgenesis in wheat (*Triticum aestivum*) anthers cultured in vitro. Sci Sin 16(2):218–225

Wedzony M (2003) Protocol for anther culture in hexaploid triticale (*Triticosecale Wittm*.). In: Maluszynski M, Kasha KJ, Forster BP, Szarejko L (eds) Doubled haploid production in crop plants: a manual. Kluwer Academic, Dordrecht, pp 123–128

Wedzony M, Marcinska I, Ponitka A, Slusarkiewicz-Jarzina A, Wozna J (1998) Production of doubled haploids in Triticale (X *Triticosecale Wittm*) by means of crosses with maize (*Zea mays*L)usingpicloramanddicamba.PlantBreed117(3):211–215.doi:10.1111/j.1439-0523.1998.tb01928.x

Wedzony M, Forster BP, Zur I, Golemiec E, Szechynska-Hebda M, Dubas E, Gotebiowska G (2009) Progress in doubled haploid technology in higher plants. In: Touraev A, Forster BP, Jain SM (eds) Advances in haploid production in higher plants. Springer, Dordrecht, pp 1–33. doi:10.1007/978-1-4020-8854-4_1

Wehmeyer N, Hernandez LD, Finkelstein RR, Vierling E (1996) Synthesis of small heat-shock proteins is part of the developmental program of late seed maturation. Plant Physiol 112(2): 747–757. doi:10.1104/pp. 112.2.747

Wojnarowiez G, Jacquard C, Devaux P, Sangwan RS, Clement C (2002) Influence of copper sulfate on anther culture in barley (*Hordeum vulgare* L.). Plant Sci 162(5):843–847. doi:10.1016/ s0168-9452(02)00036-5

Wolyn DJ, Nichols B (2003) Asparagus microspore and anther culture. In: Maluszynski M, Kasha KJ, Forster BP, Szarejko L (eds) Doubled haploid production in crop plants: a manual. Kluwer Academic, Dordrecht, pp 265–273

Xu L, Najeeb U, Tang GX, Gu HH, Zhang GQ, He Y, Zhou WJ (2007) Haploid and doubled haploid technology. In: Gupta SK (ed) Advances in botanical research: incorporating advances in plant pathology, vol 45. Advances in botanical research. Elsevier, Amsterdam, pp 181–216. doi:10.1016/s0065-2296(07)45007-8

Yao JL, Cohen D (2000) Multiple gene control of plastome-genome incompatibility and plastid DNA inheritance in interspecific hybrids of Zantedeschia. Theor Appl Genet 101(3):400–406. doi:10.1007/s001220051496

Yao QA, Simion E, William M, Krochko J, Kasha KJ (1997) Biolistic transformation of haploid isolated microspores of barley (*Hordeum vulgare* L). Genome 40(4):570–581. doi:10.1139/g97-075

Yu SM (1999) Cellular and genetic responses of plants to sugar starvation. Plant Physiol 121(3):687–693. doi:10.1104/pp. 121.3.687

Zarsky V, Rihova L, Tupy J (1990) Biochemical and cytological changes in young tobacco pollen during invitro starvation in relation to pollen embryogenesis. In: Progress in plant cellular and molecular biology, vol 9. Springer, Dordrecht

Zarsky V, Garrido D, Rihova L, Tupy J, Vicente O, Heberlebors E (1992) Derepression of the cellcycle by starvation is involved in the induction of tobacco pollen embryogenesis. Sex Plant Reprod 5(3):189–194. doi:10.1007/bf00189810

Zarsky V, Garrido D, Eller N, Tupy J, Vicente O, Schoffl F, Heberlebors E (1995) The expression of a small heat-shock gene is activated during induction of tobacco pollen embryogenesis by starvation. Plant Cell Environ 18(2):139–147. doi:10.1111/j.1365-3040.1995.tb00347.x

Zhang YL, Li DS (1984) Anther culture of monosomics in *Triticum aestivum*. Hereditas 6(3):7–10

Zhang GQ, Zhang DQ, Tang GX, He Y, Zhou WJ (2006) Plant development from microsporederived embryos in oilseed rape as affected by chilling, desiccation and cotyledon excision. Biol Plant 50(2):180–186. doi:10.1007/s10535-006-0004-6

Zhao JP, Simmonds DH, Newcomb W (1996a) High frequency production of doubled haploid plants of *Brassica napus* cv Topas derived from colchicine-induced microspore embryogenesis without heat shock. Plant Cell Rep 15(9):668–671

Zhao JP, Simmonds DH, Newcomb W (1996b) Induction of embryogenesis with colchicine instead of heat in microspores of *Brassica napus* L cv Topas. Planta 198(3):433–439. doi:10.1007/bf00620060

Zheng MY (2003) Microspore culture in wheat (*Triticum aestivum*)-doubled haploid production via inducedembryogenesis.PlantCellTissueOrganCult73(3):213–230.doi:10.1023/a:1023076213639

Zhou H (1996) Genetics of green plant regeneration from anther culture in cereals. In: Jain SM, Sopory SK, Veilleux RE (eds) In vitro haploid production in higher plants, vol 2. Kluwer Academic, Dordrecht, pp 169–187

Zhou H, Zheng Y, Konzak CF (1991) Osmotic potential of media affecting green plant percentage in wheat anther culture. Plant Cell Rep 10(2):63–66

Zhou HP, Ball ST, Konzak CF (1992) Functional-properties of ficoll and their influence on anther culture responses of wheat. Plant Cell Tissue Organ Cult 30(1):77–83. doi:10.1007/bf00040004

Zhu ZQ, Sun JS, Wang JJ (1978) Cytological investigation on androgenesis of *Triticum aestivum*. Acta Bot Sin 20(1):6–12

Zoriniants S, Tashpulatov AS, Heberle-Bors E, Touraev A (2005) The role of stress in the induction of haploid microspore embryogenesis. In: Palmer CE, Keller WA, Kasha KJ (eds) Biotechnology in agriculture and forestry, vol 56. Springer, Berlin, pp 35–52

Chapter 3
Gynogenesis: An Important Tool for Plant Breeders

Gynogenesis is the least adopted method to produce haploid plants, but it has been predominantly exploited in those crops that have shown a very little or no response to wide hybridization, microspore, or anther culture (Forster et al. 2007). Gynogenesis consists of in vitro culture of unfertilized gametes (female) such as ovaries or ovules, though occasionally complete flower buds have also been used for culture. It has been recommended that female gametes or flower buds for gynogenesis should be collected before anthesis (pollen shedding). However, the collection can be made at any time in case of a male sterile or self-incompatible species. For collection purposes, stage of microspores is an excellent indicator to identify the exact time with respect to female gametes. The procedures of surface sterilization are generally used, similar to androgenesis techniques, to sterilize/disinfect. The sterilization time and agent differ from one species to another. The donor plants grown in cold chambers or green houses mostly need less time to sterilize than plants grown under field conditions. Solid medium is the most commonly used medium to produce haploids via gynogenesis. It has been recommended to dry the ovules that need to be cultured prior to start the excision procedure. The irradiated pollen using cobalt-60 has also been used in some tree species to induce gynogenesis. The time of application and application dose have been termed as the most important factors leading towards gynogenic success. Recently, tetraploid (Germana and Chiancone 2001) and irradiated (Froelicher et al. 2007) pollens have been effectively used to induce gynogenesis in citrus.

The work on gynogenesis started in 1964 when Tulecke (1964) reported callus formation from female gametes for the first time. However, the advancement and improvement in gynogenesis was much slower than androgenesis. In barley, Noeum (1976) gave details of first haploid plant via gynogenesis by culturing ovaries. In few crop plants, such as wheat, barley, rice, and maize, doubled haploidy via gynogenesis is possible (Gaj 1998; Sibi et al. 2001; Tang et al. 2006; Zhou and Yang 1981), but androgenesis is a method of choice due to the reason of low embryo

M. Asif, *Progress and Opportunities of Doubled Haploid Production*,
SpringerBriefs in Plant Science 6, DOI 10.1007/978-3-319-00732-8_3,
© Springer International Publishing Switzerland 2013

production and limited number of cells to manipulate in gynogenesis. Gynogenesis has been effectively used in sugar beet (*Beta vulgaris*) and onion (*Allium cepa*) as described by Gurel et al. (2000) and Luthar and Bohanec (1999), respectively. The frequency of haploid production via gynogenesis in angiosperms, cucumber (Gemes-Juhasz et al. 2002), sweet potato (Kobayashi et al. 1993), and in some trees (Forster et al. 2007) has shown promising results. In sugar beet and onion, gynogenesis has been achieved by culturing female gametophyte followed by their regeneration. There is no need to apply pretreatment in case of onion, but a cold shock (8 °C for 7 days) to floral organs in combination with high temperature treatment at 30 °C during culturing is needed in sugar beet to obtain desired results (Michalik et al. 2000; Wremerth and Levall 2003). Major factors that influence gynogenesis include genotype, developmental stage of female gametophyte or embryo sac, pretreatment, composition of media, and induction/cultural conditions.

3.1   Genotype

Haploid production via gynogenesis is dependent on the type of genotype being used and is highly variable from one species to another. The induction response (no. of embryos) also depends on donor plant's growth and developmental conditions and quality of female gametophytes at the time of induction. In rice (Rongbai et al. 1998), onion (Bohanec et al. 1995), and wheat (Mdarhri-Alaoui et al. 1998), genotypic variations with respect to gynogenic success of haploid production have been well documented. The variable gynogenic induction frequency has been illustrated in onion genotypes of various origins that ranged from 0 to 10 % among 10 Polish cultivars/varieties (Javornik et al. 1998), 0 to 17 % in 22 European accessions (Geoffriau et al. 1997), and 0 to 22 % among 39 Japanese and American accessions (Bohanec and Jakse 1999). Furthermore, open pollinated cultivars of *Allium cepa* showed low response to gynogenesis than inbred lines and F_1 hybrids (Bohanec and Jakse 1999). A similar genotypic variability was also noted in potato and squash cultivars that ranged from 11 to 60 (Kobayashi et al. 1993) and 0 to 49 % (Shalaby 2007), respectively. Higher embryo yield was noted in sugar beet when floral organs were collected from donor plants grown in glass cabinets/greenhouses compared to field grown plants (Lux et al. 1990). In the same way, florets that developed first on lateral branches yielded higher number of embryos as compared to florets emerged at the later stages i.e. florets developed at main stem apex (D'Halluin and Keimer 1986a, b). The planting/sowing time also affected embryoid response and it has been stated that summer planting favored embryo production in *Beta* sp., whereas autumn sowing yielded higher embryos in *Gerbera* sp. (Cappadocia and Ahmim 1988; Cappadocia et al. 1988; Doctrinal et al. 1989; Lux et al. 1990). In onion, female gametophytes collected from very large and small flowers showed a very little response to gynogenic induction than medium ones exhibiting two to four nucleate embryo sacs (Musial et al. 2005).

3.2   Developmental Stage of Female Gametophyte

The exact stage of female gametophytic (ovule or ovary) regarding its collection for induction is hard to detect. The induction of ovules/ovaries is done on the basis of microspore stage or days after anthesis; however, in few cases, flower bud's developmental stage or direct observation of ovaries/ovules has also been performed. Nearly mature or mature embryo sac is considered as a good sign to commence the process of gynogenesis (Gemes-Juhasz et al. 2002; Keller and Korzun 1996); however, a well-trained and skilled embryologist is needed to determine the exact stage of embryo sac. Yang et al. (1986) confirmed that in sunflower, a well mature embryo sac is developed just 2–3 days before anthesis, whereas an excellent response has been obtained from unfertilized ovules of "Kyoho" grape wine when they were collected/excised, 19–20 days before anthesis (Nakajima et al. 2000). In the same way, ovaries harvested 1 day prior to anthesis in *Cucurbita pepo* resulted maximum embryos (Metwally et al. 1998a, b), whereas similar findings have been reported in cucumber when excision was done only 6 h prior to anthesis (Gemes-Juhasz et al. 2002). The embryo sac consists of egg cell, two polar nuclei, antipodal cells, and synergids. In most cases, egg cell give rise to haploid embryo but sometimes production of an embryo from synergid or antipodal cells has also been documented as in the case of rice (Zhou et al. 1986) and barley (Noeum 1976). In saffron, the excision time of ovaries has been linked to stigma development. Ovaries with yellow stigma are considered one the most responsive stage for gynogenic excision (Bhagyalakshmi 1999).

3.3   Pretreatment

The floral organs are pretreated in few species which is also considered as an important factor to stimulate the process of sporophytic development in female gametophytes. Similar to androgenic methods, starvation, heat, and/or cold shocks are given alone or in combination with each other to induce stress conditions. However, duration, time, type, and level of pretreatment vary considerably from one species to another. In rice, cold treatment at 8 °C for 6–14 days enhanced embryogenic response to a greater extent (Rongbai et al. 1998) and similar effect of cold treatment has been seen in sugar beet, wheat, and *Salvia sclarea* (Bugara and Rusina 1989a, b; Gurel et al. 2000; Sibi et al. 2001). Yang et al. (1986) reported that a rapid cold shock for 1–2 days at 4 °C also improved embryogenic efficiency in *Helianthus* sp. On the other hand, heat treatment for 2–4 days at 33 °C seemed to be efficient to promote sporophytic development of female gametophyte in *Picea sitchensis* (Baldursson et al. 1993). Gemes-Juhasz et al. (2002) described that heat shock (32 °C) is also effective in cucumber to obtain an effective gynogenic response provided that heat treatment is given during cultural/induction phase. In few species such as *Cucurbita pepo* (Metwally et al. 1998a, b), niger (Bhat and Murthy 2007), and rice (Zhou et al. 1986), there is no need to apply any pretreatment and heat/cold shocks have shown detrimental effects in these species. In the same way, high illumination favors onion gynogenesis (Puddephat et al. 1999), whereas dark incubation during induction phase is required in saffron (Bhagyalakshmi 1999) and cucumber (Gemes-Juhasz et al. 2002).

3.4 Composition of Media

The composition of media is also a critical factor that affects success rate of gynogenesis to a considerable degree. Media constituents differ not only for regeneration and induction phases but also among crops. The media of regeneration phase require growth hormones/regulators to promote growth, whereas low concentration or sometime even no growth regulators are needed for induction medium. MS, Millers, N6, and B5 with minor changes/modifications in the sources of growth regulators, carbohydrates, and nitrogen are among the most widely adapted media.

Sucrose is the most extensively used carbohydrate and its concentration in media vary from 58 to 348 mM or 2 to 12 % (Juhasz et al. 1997; Mdarhri-Alaoui et al. 1998). The sucrose in a concentration of 6 % in the media enhanced number of embryos and hampered somatic tissue growth in wheat gynogenesis (Mukhambetzhanov 1997), but its high concentration believed to be beneficial in carnation and has an adverse effect on the frequency of embryos in squash (Sato et al. 2000). On the other hand, in few species, maltose has been exploited/used rather than sucrose (Cordewener et al. 1995). For gynogenic haploid production, Cytokinin and Auxin have been mainly used in various crop species, but it has been observed that novel polyamines gave much better response as compared to growth hormones/regulators in onion and their substitution has indicated better results (Martinez et al. 2000). Several other growth hormones such as indole-3-acetic acid (IAA) in onion (Bohanec et al. 1995) and carrot (Kielkowska and Adamus 2010), naphthalene acetic acid (NAA) in rice (Rongbai et al. 1998), coconut water in barley (Castillo and Cistue 1993), dimethyl sulfoxide (DMSO) in rice (Rongbai et al. 1998), and Thidiazuron (TDZ) in cucumber (Diao et al. 2009) alone or in combination with each other have given promising gynogenic results. In wheat, solid medium is preferred over liquid because it improves callus growth (Gusakovskaya and Najar 1994).

A significant amount of research work has been conducted in the area of doubled haploidy via gynogenesis over the past few decades. The research papers have addressed various aspects of gynogenesis; however, the major focal point has been to improve the methodology using responsive genotypes and altering donor plant's conditions, pretreatments, and media composition. Besides many recent gynogenesis publications, a very little information exists with respect to molecular and genetics of gynogenesis in crop plants. As described earlier, gynogenesis is considered a method of choice where other methods of doubled haploidy are not available or species are irresponsive to other methods. This is especially true in case of onion, melon, and sugar beet where significant achievements have been made with gynogenesis because androgenesis via anther culture or IMC is not successful. Furthermore, gynogenesis has also proved to be beneficial in case of male sterile plants e.g. haploids via gynogenesis have been effectively developed in photosensitive male sterile line in rice by culturing unfertilized ovaries (Cai et al. 1988). The problem of albinism in cereals can also be tackled with gynogenesis. In petunia, the percentage of nonhaploids were more than haploids using androgenesis (anther culture) as a method of haploid production, but DeVerna and Collins (1984) reported that 93 % plants were haploids when produced through gynogenesis. Similar findings in rice with the use of gynogenesis have been reported. It is suggested that future gynogenesis work should be conducted to improve our knowledge to track gynogenic sporophytic pathway and to identify new QTLs or genes associated with it. The development and identification of genetic/molecular markers associated with higher frequency of embryoids and green plants through gynogenesis will definitely improve gynogenic doubled haploidy in crop plants to a considerable degree.

References

Baldursson S, Norgaard JV, Krogstrup P (1993) Factors influencing haploid callus initiation and proliferation in megagametophyte cultures of sitka spruce (*Picea sitchensis*). Silvae Genet 42(2–3):79–86

Bhagyalakshmi N (1999) Factors influencing direct shoot regeneration from ovary explants of saffron. Plant Cell Tissue Organ Cult 58(3):205–211. doi:10.1023/a:1006398205936

Bhat JG, Murthy HN (2007) Factors affecting in-vitro gynogenic haploid production in niger (*Guizotia abyssinica* (L. f.) Cass.). Plant Growth Regul 52(3):241–248. doi:10.1007/s10725-007-9196-9

Bohanec B, Jakse M (1999) Variations in gynogenic response among long-day onion (*Allium cepa* L.) accessions. Plant Cell Rep 18(9):737–742

Bohanec B, Jakse M, Ihan A, Javornik B (1995) Studies of gynogenesis in onion (*Allium cepa* L) – induction procedures and genetic-analysis of regenerants. Plant Sci 104(2):215–224. doi:10.1016/0168-9452(94)04030-k

Bugara AM, Rusina LV (1989a) Haploid callus formation in the culture of unfertilized ovules of clary sage. Fiziologiya i Biokhimiya Kulturnykh Rastenii 20(6):554–560

Bugara AM, Rusina LV (1989b) Haploid callusogenesis in the culture of unfertilized ovules of *Salvia sclarea*. Fiziologiya I Biokhimiya Kulturnykh Rastenii 21(6):554–560

Cai DT, Chen DT, Zhu H, Jin Y (1988) In vitro production of haploid plantlets from the unfertilized ovaries and anthers of Hubei photosynthetic genicmale sterile rice (HPGMR). Acta Biol Exp Sin 21:401–407

Cappadocia M, Ahmim M (1988) Comparison of 2 culture methods for the production of haploids by anther culture in *Solanum chacoense*. Can J Bot 66(5):1003–1005

Cappadocia M, Chretien L, Laublin G (1988) Production of haploids in *Gerbera-jamesonii* via ovule culture – influence of fall versus spring sampling on callus formation and shoot regeneration. Can J Bot 66(6):1107–1110

Castillo AM, Cistue L (1993) Production of gynogenic haploids of *Hordeum vulgare* L. Plant Cell Rep 12(3):139–143

Cordewener JHG, Hause G, Gorgen E, Busink R, Hause B, Dons HJM, Vanlammeren AAM, Campagne MMV, Pechan P (1995) Changes in synthesis and localization of members of the 70-kDa class of heat-shock proteins accompany the induction of embryogenesis in *Brassica napus* L microspores. Planta 196(4):747–755. doi:10.1007/bf01106770

D'Halluin K, Keimer B (1986a) Production of haploid sugarbeets (*Beta vulgaris* L.) by ovule culture. Genetic manipulation in plant breeding. In: Proceedings of international symposium organized by Eucarpia, 8–13 Sept 1985, Berlin (West), Germany, pp 307–309

D'Halluin K, Keimer B (1986b) Production of haploid sugarbeets *Beta vulgaris* L. By ovule culture. In: Horn W, Jensen CJ, Odenbach W, Schieder O (eds) Genetic manipulation in plant breeding. De Gruyter, Berlin, pp 307–310

DeVerna JW, Collins GB (1984) Maternal haploids of *Petunia axillaris* (Lam.) B.S.P. via culture of placenta attached ovules. Theor Appl Genet 69(2):187–192. doi:10.1007/BF00272893

Diao W-P, Jia Y-Y, Song H, Zhang X-Q, Lou Q-F, Chen J-F (2009) Efficient embryo induction in cucumber ovary culture and homozygous identification of the regenetants using SSR markers. Sci Hortic 119(3):246–251. doi:10.1016/j.scienta.2008.08.016

Doctrinal M, Sangwan RS, Sangwannorreel BS (1989) Invitro gynogenesis in *Beta vulgaris* L – effects of plant-growth regulators, temperature, genotypes and season. Plant Cell Tissue Organ Cult 17(1):1–12

Forster BP, Heberle-Bors E, Kasha KJ, Touraev A (2007) The resurgence of haploids in higher plants. Trends Plant Sci 12(8):368–375. doi:10.1016/j.tplants.2007.06.007

Froelicher Y, Bassene J-B, Jedidi-Neji E, Dambier D, Morillon R, Bernardini G, Costantino G, Ollitrault P (2007) Induced parthenogenesis in mandarin for haploid production: induction procedures and genetic analysis of plantlets. Plant Cell Rep 26(7):937–944. doi:10.1007/s00299-007-0314-y

Gaj M (1998) Estimation of parthenogenesis frequency in genotypes of *Hordeum vulgare* (L.) by auxin test. J Appl Genet 39A:98

Gemes-Juhasz A, Balogh P, Ferenczy A, Kristof Z (2002) Effect of optimal stage of female gametophyte and heat treatment on in vitro gynogenesis induction in cucumber (*Cucumis sativus* L.). Plant Cell Rep 21(2):105–111. doi:10.1007/s00299-002-0482-8

Geoffriau E, Kahane R, Rancillac M (1997) Variation of gynogenesis ability in onion (*Allium cepa* L). Euphytica 94(1):37–44. doi:10.1023/a:1002949606450

Germana MA, Chiancone B (2001) Gynogenetic haploids of Citrus after in vitro pollination with triploid pollen grains. Plant Cell Tissue Organ Cult 66(1):59–66. doi:10.1023/a:1010627310808

Gurel S, Gurel E, Kaya Z (2000) Doubled haploid plant production from unpollinated ovules of sugar beet (*Beta vulgaris* L.). Plant Cell Rep 19(12):1155–1159

Gusakovskaya MA, Najar MA (1994) Gynogenesis in cultures of ovaries and ovules of *Triticum aestivum* L. (Poaceae). Botanicheskii Zhurnal 79(1):70–79

Javornik B, Bohanec B, Campion B (1998) Second cycle gynogenesis in onion, *Allium cepa* L, and genetic analysis of the plants. Plant Breed 117(3):275–278. doi:10.1111/j.1439-0523.1998.tb01939.x

Juhasz AG, Simon-Sarkadi L, Velich I, Varro P (1997) Studies of non-ionic osmotic stress on bean (*Phaseolus vulgaris* L) callus and seedlings cultures. Acta Hortic 447:455–456

Keller ERJ, Korzun L (1996) Haploidy in onion (*Allium cepa* L.) and other Allium species. In: Jain SM, Sopory SK, Veilleux RE (eds) In vitro haploid production in higher plants, vol 3, Important selected plants. Kluwer, Dordrecht, pp 51–75

Kielkowska A, Adamus A (2010) In vitro culture of unfertilized ovules in carrot (*Daucus carota* L.). Plant Cell Tissue Organ Cult 102(3):309–319. doi:10.1007/s11240-010-9735-3

Kobayashi RS, Sinden SL, Bouwkamp JC (1993) Ovule culture of sweet-potato (*Ipomoea batatas*) and closely related species. Plant Cell Tissue Organ Cult 32(1):77–82. doi:10.1007/bf00040119

Luthar Z, Bohanec B (1999) Induction of direct somatic organogenesis in onion (*Allium cepa* L.) using a two-step flower or ovary culture. Plant Cell Rep 18(10):797–802

Lux H, Herrmann L, Wetzel C (1990) Production of haploid sugar-beet (*Beta vulgaris* L) by culturing unpollinated ovules. Plant Breed 104(3):177–183. doi:10.1111/j.1439-0523.1990.tb00420.x

Martinez LE, Aguero CB, Lopez ME, Galmarini CR (2000) Improvement of in vitro gynogenesis induction in onion (*Allium cepa* L.) using polyamines. Plant Sci 156(2):221–226. doi:10.1016/s0168-9452(00)00263-6

Mdarhri-Alaoui M, Saidi N, Chlyah A, Chlyah H (1998) Green haploid plant formation in durum-wheat through in vitro gynogenesis. CR Acad Sci Ser III Life 32:25–30

Metwally EI, Moustafa SA, El-Sawy BI, Haroun SA, Shalaby TA (1998a) Production of haploid plants from in vitro culture of unpollinated ovules of *Cucurbita pepo*. Plant Cell Tissue Organ Cult 52(3):117–121. doi:10.1023/a:1005948809825

Metwally EI, Moustafa SA, El-Sawy BI, Shalaby TA (1998b) Haploid plantlets derived by anther culture of *Cucurbita pepo*. Plant Cell Tissue Organ Cult 52(3):171–176. doi:10.1023/a:1005908326663

Michalik B, Adamus A, Nowak E (2000) Gynogenesis in Polish onion cultivars. J Plant Physiol 156(2):211–216

Mukhambetzhanov SK (1997) Culture of nonfertilized female gametophytes in vitro. Plant Cell Tissue Organ Cult 48(2):111–119. doi:10.1023/a:1005838113788

Musial K, Bohanec B, Jakse M, Przywara L (2005) The development of onion (*Allium cepa* L.) embryo sacs in vitro and gynogenesis induction in relation to flower size. In Vitro Cell Dev Biol Plant 41(4):446–452. doi:10.1079/ivp2005645

Nakajima I, Kobayashi S, Nakamura Y (2000) Embryogenic callus induction and plant regeneration from unfertilized ovule of "Kyoho" grape. J Jpn Soc Hortic Sci 69(2):186–188

Noeum LHS (1976) Non-fertilized ovaries of *Hordeum vulgare* L. cultured in vitro (Haploides d'Hordeum vulgare L. par culture in vitro d'ovaires non fecondes.). Annales de l'Amelioration des Plantes 26(4):751–754

Puddephat IJ, Robinson HT, Smith BM, Lynn J (1999) Influence of stock plant pretreatment on gynogenic embryo induction from flower buds of onion. Plant Cell Tissue Organ Cult 57(2):145–148. doi:10.1023/a:1006312614874

Rongbai L, Pandey MP, Pandey SK, Dwivedi DK, Ashima (1998) Exploiting the in vitro ovary culture technique to breed rice hybrids. Int Rice Res Notes 23(1):14

Sato S, Katoh N, Yoshida H, Iwai S, Hagimori M (2000) Production of doubled haploid plants of carnation (*Dianthus caryophyllus* L.) by pseudofertilized ovule culture. Sci Hortic 83(3–4): 301–310. doi:10.1016/s0304-4238(99)00090-4

Shalaby TA (2007) Factors affecting haploid induction through in vitro gynogenesis in summer squash (*Cucurbita pepo* L.). Sci Hortic 115(1):1–6. doi:10.1016/j.scienta.2007.07.008

Sibi ML, Kobaissi A, Shekafandeh A (2001) Green haploid plants from unpollinated ovary culture in tetraploid wheat (*Triticum durum* Defs.). Euphytica 122(2):351–359. doi:10.1023/a:1012991325228

Tang F, Tao Y, Zhao T, Wang G (2006) In vitro production of haploid and doubled haploid plants from pollinated ovaries of maize (*Zea mays*). Plant Cell Tissue Organ Cult 84(2):210–214. doi:10.1007/s11240-005-9017-7

Tulecke W (1964) Haploid tissue culture from female gametophyte of *Ginkgo biloba* L. Nature 203(494):94. doi:10.1038/203094a0

Wremerth WE, Levall M (2003) Doubled haploid production of sugar beet (*Beta vulgaris* L.). In: Maluszynski M, Kasha KJ, Forster BP, Szarejko I (eds) Doubled haploid production in crop plants: a manual. Kluwer, Dordrecht, pp 255–265

Yang HY, Zhou C, Cai D, Hua Y, Yan W, Chen X (1986) In vitro culture of unfertilized ovules in *Helianthus annuus* L. In: Hu H, Yang HY (eds) Haploids of higher plant: in vitro. Springer, Berlin, pp 182–191

Zhou C, Yang HY (1981) In vitro embryogenesis in unfertilized embryo sacs of Oryza sativa L. Acta Bot Sin 23:176–180

Zhou C, Yang HY, Tian HQ, Liu ZL, Yan H (1986) In vitro culture of unpollinated ovaries in *Oryza sativa* L. In: Hu H, Yang HY (eds) Haploids of higher plants in vitro. China Academic, Beijing, pp 165–181

Chapter 4
Parthenogenesis

Haploid plant production via parthenogenesis involves culturing egg cell in the embryo sac without any involvement of sperm nuclei. Palmer and Keller (2005) differentiated parthenogenesis from gynogenesis whereby the former involves "the normal development of endosperm and embryo formation occurs in vivo," whereas in the case of gynogenesis, endosperm degenerates with the passage of time and there is a need to rescue the embryo in the laboratory conditions. Chase (1949) produced doubled haploids in maize via parthenogenesis and exploited these haploids in his breeding program. He used a color genetic marker (dominant purple) in the pollinator to distinguish haploids (colorless) from diploids that followed chromosome doubling using a colchicine injection in the scutellar node of maize haploid plants. Since parthenogenesis occurs rarely in nature, it is, therefore, difficult to distinguish between diploids and haploids. Thus, genetic markers are usually used in the pollinators for selection purposes as described by Bordes et al. (1997) in apple. The induction of parthenogenesis is usually brought out by using irradiated pollen, heat treatment, and gametocidal chemicals. The pollen was successfully treated with heat to produce haploids in maize by Mathur et al. (1980). The use of chemicals to treat pollen is also common and it has also been applied in maize (Deanon 1957) and brassica (Kitani 1994). As described earlier, one of the best example of parthenogenesis is the production of haploid plants in cultivated tetraploid potato (*Solanum tuberosum*) species by crossing it with diploid *S. phureja* (pollen donor). The genetic control of parthenogenesis has been identified in maize and barley where indeterminate gametophyte (*ig*) and *hap* initiator genes are capable to induce parthenogenesis in maize (Kermicle 1969) and barley (Hagberg and Hagberg 1980), respectively. The frequency of haploids occurrence in nature is very low. An auxin test has been identified to estimate the frequency of parthenogenic haploids (Mazzucato et al. 1996). However, the use of inducer and marker genes will definitely improve the recovery of haploids to be used or exploit this method to breed genotypes on a larger and commercial scale.

M. Asif, *Progress and Opportunities of Doubled Haploid Production*,
SpringerBriefs in Plant Science 6, DOI 10.1007/978-3-319-00732-8_4,
© Springer International Publishing Switzerland 2013

References

Bordes J, deVaulx RD, Lapierre A, Pollacsek M (1997) Haplodiploidization of maize (*Zea mays* L) through induced gynogenesis assisted by glossy markers and its use in breeding. Agronomie 17(5):291–297. doi:10.1051/agro:19970504

Chase SS (1949) Monoploid frequencies in a commercial double cross hybrid maize, and in its component single cross hybrids and inbred lines. Genetics 34(3):328–332

Deanon JR (1957) Treatment of sweet corn silks with maleic hydrazide and colchicine as means of increasing the frequency of monoploids. Philipp Agric 41(7):364–377

Hagberg A, Hagberg G (1980) High-frequency of spontaneous haploids in the progeny of an induced mutation in barley. Hereditas 93(2):341–343

Kermicle JL (1969) Androgenesis conditioned by a mutation in maize. Science 166(3911):1422. doi:10.1126/science.166.3911.1422

Kitani Y (1994) Induction of parthenogenetic haploid plants with brassinolide. Jpn J Genet 69(1):35–39. doi:10.1266/jjg.69.35

Mathur DS, Aman MA, Sarkar KR (1980) Induction of maternal haploids in maize through heat-treatment of pollen. Curr Sci 49(19):744–746

Mazzucato A, denNijs APM, Falcinelli M (1996) Estimation of parthenogenesis frequency in Kentucky bluegrass with auxin-induced parthenocarpic seeds. Crop Sci 36(1):9–16

Palmer CE, Keller WA (2005) Overview of haploidy. In: Palmer CE, Keller WA, Kasha KJ (eds) Haploids in crop improvement II. Springer, Berlin, pp 3–9

Chapter 5
Applications and Uses of Haploids

The use of doubled haploidy has a great impact on plant breeding programs through-out the world especially to achieve food security on sustainable basis, which is a serious concern not only to developing countries but also to developed nations. There is a sharp increase in in-vitro studies to improve DH production and method-ology/protocol (Table 5.1). Germana (2011) reported that 50 % of the barley culti-vars currently grown in Europe have been developed by means of DH technologies. According to Faostat (2010), Canada is world's 2nd biggest wheat exporter after USA, producing around 23 metric tons wheat every year. In Canada, wheat has been classified into nine different classes, out of which Canada Western Red Spring (CWRS) is the biggest class. It has been reported that during 2007, 3 out of 5 most grown cultivars of CWRS were developed through doubled haploid technology. Furthermore, cultivar "AC Andrew" is being grown on an area of 99 % under Canada Western Soft White Spring (CWSWS) class and it was also developed via anther culture (Dunwell 2010). Mapping populations are also being developed through DH that offers great potential for genetic studies, molecular analysis, and to map DNA markers. It was estimated that around 290 cultivars grown in various parts of the world during 2005 were developed through DH techniques (http://www. scri.ac.uk/assoc/COST851/DHTable2005.xls). Emerging uses of DH technology in plant breeding and crop improvement programs have been discussed as follows.

5.1 Homozygosity

Homozygosity in crop plants can be fixed only in one generation with the use of doubled haploidy whereas in normal conventional breeding, 6–7 years of continu-ous selfing is required. Therefore, doubled haploidy allows plant breeders and geneticists to release a cultivar in 6–7 rather than 10–12 years and this is especially true for species where the plants breeders are getting only one crop per year like winter wheat. Doubled haploid lines are homozygous and homogeneous; therefore,

M. Asif, *Progress and Opportunities of Doubled Haploid Production*,
SpringerBriefs in Plant Science 6, DOI 10.1007/978-3-319-00732-8_5,
© Springer International Publishing Switzerland 2013

Table 5.1 Studies showing improvement/successful doubled haploid production in various species during last 5 years

Crop	Method	Reference
Rice	AC	Bagheri et al. (2009), Chaitali and Singh (2011), Chen et al. (2010b), Herath and Bandara (2011), Yang et al. (2011), Yuan et al. (2011)
Wheat	AC	Barakat et al. (2012), Broughton (2008, 2011), Danci et al. (2011), El-Hennawy et al. (2011), Hassawi et al. (2012), Islam (2010c), Ljevnaic et al. (2009), Mozgova et al. (2012), Redha and Suleman (2011), Redha and Talaat (2008), Santra et al. (2012), Song et al. (2012), Soriano et al. (2008), Weyen et al. (2012), Zhao et al. (2012)
Barley	AC	Belinskaya (2010), Bilynska and Dulnyev (2012), Dyulgerova et al. (2010)
Triticale	AC	Krzewska et al. (2012), Mozgova et al. (2012), Ponitka and Slusarkiewicz-Jarzina (2011), Sun et al. (2009)
Oat	AC	Kiviharju (2009), Marcinska et al. (2013)
Maize	AC	Jaeger et al. (2010), Marcinska et al. (2013), Obert et al. (2009)
Pea	AC	Bobkov (2010)
Linseed	AC	Burbulis and Blinstrubiene (2011), Burbulis et al. (2012)
Pear	AC	Tang et al. (2009, 2012)
Grapevine	AC	Alavijeh et al. (2012)
Eggplant	AC	Basay et al. (2011), Corral-Martinez and Segui-Simarro (2012), Ellaltoglu et al. (2012), Salas et al. (2011, 2012)
Citrus	AC	Benelli et al. (2010)
Flax	AC	Burbulis et al. (2009), Mankowska et al. (2011)
Tobacco	AC	Chen et al. (2010a)
Asparagus	AC	Ercan and Sensoy (2012)
Alfalfa	AC	Ma and Zhang (2008), Geng et al. (2010)
Pepper	AC	Grozeva et al. (2009), Irikova et al. (2011a, b), Koleva-Gudeva et al. (2009), Li et al. (2012), Ochoa-Alejo (2012), Olszewska et al. (2011), Supena and Custers (2011), Taskin et al. (2011), Zhao et al. (2010)
Carrot	AC	Gorecka et al. (2009), Kiszczak et al. (2011), Krystyna et al. (2010), Szafranska et al. (2011)
Brassica sp.	AC	Sayem et al. (2010)
Cotton	AC	Shabana et al. (2010)
Tomato	AC	Motallebi-Azar (2010), Motallebi-Azar and Panahandeh (2010), Shere and Dhage (2009)
Wheat	IMC	Cistue et al. (2009), Islam (2010a, b), Lantos et al. (2009), Ren et al. (2010), Santra et al. (2012), Shirdelmoghanloo et al. (2009)
Barley	IMC	Esteves et al. (2010), Jacquard et al. (2009a, b), Rodriguez-Serrano et al. (2012)
Oat	IMC	De Cesaro et al. (2009), Sidhu and Davies (2009)
Brassica	IMC	Barbulescu et al. (2011), Belmonte et al. (2010, 2011), Bhowmik et al. (2011), Ghazanfari et al. (2012), Jo et al. (2012), Kim and Lee (2012), Lee et al. (2011), Malik and Krochko (2009), Mohammadi et al. (2012), Na et al. (2009, 2011a, b), Nelson et al. (2009a, b), Prem et al. (2012), Segui-Simarro et al. (2011), Takahashi et al. (2011, 2012), Takahira et al. (2011), Wan et al. (2011), Wang et al. (2011), Wen et al. (2010), Winarto and da Silva (2011), Yadollahi et al. (2011), Yuan et al. (2011, 2012), Zeng et al. (2010), Zhang et al. (2011, 2012)

(continued)

Table 5.1 (continued)

Crop	Method	Reference
Pepper	IMC	Kim et al. (2013), Lantos et al. (2009, 2012), Supena and Custers (2011), Taskin et al. (2011)
Eggplant	IMC	Corral-Martinez and Segui-Simarro (2012)
Chickpea	IMC	Croser et al. (2011)
Maize	IMC	de Moraes et al. (2008), Obert et al. (2009), Testillano et al. (2010)
Triticale	IMC	Dubas et al. (2010), Zur et al. (2009)
Carrot	IMC	Gorecka et al. (2010)
Cotton	IMC	Poon et al. (2012)
Caraway	IMC	Smykalova et al. (2012)
Rye	IMC	Targonska et al. (2013)
Wheat	Wide crossing	Chen et al. (2011), Gu et al. (2008), Khan and Ahmad (2011), Kour et al. (2008), Prodanovic et al. (2008), Usha and Khanna (2010), Zhang et al. (2011), Bouatrous et al. (2010)
Barley	Wide crossing	Houben et al. (2011)
Oat	Wide crossing	Marcinska et al. (2013)
Potato	Wide crossing	Weber et al. (2012)
Chickpea	Wide crossing	Clarke et al. (2011)
Citrus	Wide crossing	Yahata et al. (2010)
Niger	Gynogenesis	Bhat and Murthy (2008)
Onion	Gynogenesis	Forodi et al. (2009), Hyde et al. (2012), Liu et al. (2010)
Gentians	Gynogenesis	Doi et al. (2011)
Artichoke	Gynogenesis	Guerrand et al. (2012)
Carrot	Gynogenesis	Kielkowska and Adamus (2010)
Gentiana triflora	Gynogenesis	Pathirana et al. (2011)
Cucurbita	Gynogenesis	Rakha et al. (2012)
Cucumber	Gynogenesis	Suprunova and Shmykova (2008), Diao et al. (2009)
Beta vulgaris	Gynogenesis	Tomaszewska-Sowa (2010)
Cotton	Gynogenesis	Kantartzi and Roupakias (2009)
Pumpkins	Parthenogenesis	Berber et al. (2012)
Mandarin	Parthenogenesis	Froelicher et al. (2007)
Snapmelon	Parthenogenesis	Godbole and Murthy (2012)
Walnut	Parthenogenesis	Grouh et al. (2011)
Maize	Parthenogenesis	Liu et al. (2009), Tyrnov and Smolkina (2011), Wang et al. (2010), Wei et al. (2010)
Pilosella rubra	Parthenogenesis	Rosenbaumova et al. (2012)
Chinese chive	Parthenogenesis	Yamashita et al. (2012)
Cucumber	Parthenogenesis	Lotfi and Salehi (2008)

AC Anther culture, *IMC* Isolated Microspore Culture

these lines are ideal in those species where high level of purity is required. Moreover, cost related to phenotypic selection, space and time can be dramatically reduced by using DH technology because selection is performed in true breeding (homogenous) progenies rather than the segregating populations. Eder and Chalyk (2002) pointed out that DH provide a mean of natural selection in maize where haploids plants with deleterious/harmful genes die at a very early stage of their life or they are too weak, sterile and seed setting is not done. Sometime gametophytic embryogenesis leads to an enhanced gene expression of some traits that otherwise is not possible to exploit in conventional breeding due to a control by recessive alleles in disomic state.

5.2   Genomics

Doubled haploid populations are very useful for genetic studies. Quantitative trait loci for many agronomic and quality traits have been identified in many crops using DH lines. The discovery of molecular markers has facilitated QTL analysis to a greater extent using populations like F_2, backcross, and Recombinant Inbred Lines (RILs), but studies with F_2 or backcross populations cannot be repeated and RILs takes longer time to develop due to several cycles (at least six) of selfing. Thus, populations can be quickly generated/developed through DH technology and such populations are also immortal. The fixed genetic structure of DH population provides a valuable source that permits breeders to repeat studies across various environments, allowing them to find interaction of QTL with the environments and measure the exact phenotypic expression of a particular QTL in a given environment. Recently, a large number of studies have employed DH lines to identify QTLs and develop genetic maps that include at least 100 DH populations in four cereal species: wheat, barley, rice, and maize. Few recent examples include Fusarium Head Blight (FHB) resistance in barley (Ma et al. 2000) and wheat (Suzuki et al. 2012), and plant height, flowering time (Heidari et al. 2012), photoperiod (Sourdille et al. 2000), *Septoria tritici* blotch resistance (Kelm et al. 2012), grain yield (Kuchel et al. 2007a, b), and yield components (Cuthbert et al. 2008) in wheat. Chu et al. (2008) used DH population and constructed a genetic map in wheat that consisted of 632 markers. The distorted segregation has been reported in DH populations derived through anther culture in rice but the percentage of markers having distortion was same in F_2-derived population (Yamagishi et al. 1998) and distorted segregation has also been observed in RILs population derived through Single Seed Descent (SSD) (Bjornstad et al. 1993), thereby not limiting the role of DH populations in genetics studies. He et al. (2001) compared molecular marker segregation in populations derived through anther culture (DH) and SSD (RIL). They found a distorted segregation of 27.3 % and 18.2 % in RIL and DH populations, respectively. The phenotypic evaluation of three DH populations to their respective RIL populations also showed no significant differences, and Courtois (1993) urged that both approaches are equally effective in developing new cultivars.

5.3   Mutation

Induced mutations have been used to improve traits and to create genetic variability. Maluszynski and Ahloowalia (2000) reported that 70 % of mutant varieties have been directly released for commercial cultivation without any further improvement through breeding and the rest of 30 % mutants served as parents to obtain desirable alleles. Therefore, mutation breeding is an integral part of the conventional breeding and is more useful strategy where desired combination of alleles cannot be incorporated with conventional breeding. Haploid cells offer an excellent opportunity for artificial mutation, and mutants can be easily detected/selected. In most cases, chemical mutagens, gamma, ultraviolet (UV), and X-rays have been successfully used to induce mutation in haploid cells due to their uniformity and abundance. The optimum time for mutagenesis has been described as 16–24 h after induction of microspores in the culture medium (Huang 1992) but the efficiency varies among species and mainly dependent on the type and duration of mutagen treatment. It is desirable to apply mutagens before the start of first nuclear division to avoid any heterozygosity and chimerism. Castillo et al. (2001) found that an application of sodium azide (10^{-5} to 10^{-4} M) to barley microspores for 1 h yielded 8.6–15.6 % mutants. In similar studies, ethyl methane sulfonic acid (Lantos et al. 2009) and gamma radiations (Chen et al. 2001) were used to induce mutation in rice anthers right after isolation that lead to produce stable mutants. Microspore mutagenesis have been successfully used in brassica species to develop cultivars resistant to herbicides (Swanson et al. 1989), *Alternaria brassicicola* (Ahmad et al. 1991), high oleic acid and reduced linoleic and fatty acid (Kott 1996) and modifications in erucic acid (Barro et al. 2001).

5.4 Genetic Transformation

The haploid cells during DH production act as an ideal target for genetic transformation. Transformation can be done on unicellular microspores and haploid embryos that will result in a rapid recovery of transgenics with fixed homozygosity. During embryogenesis, transformation can be performed with already established methods like microinjection, agrobacterium-mediated DNA delivery, particle bombardment, and electroporation (Touraev et al. 2001) but IMC is preferred over other methods (like gynogenesis and parthenogenesis) due to high efficiency of gene introduction. Kasha et al. (2001) urge that gene incorporation in cereals should be done prior to nuclear fusion to obtain homozygous transgenic DH plants because spontaneous chromosome doubling occurs just after first nuclear division at nuclear fusion. The frequency of transgenics using agrobacterium-mediated transformation during embryogenesis is very low (Dormann et al. 1995; Huang 1992) and the results are also nonreproducible. However, particle bombardment is one of the best methods used for microspores transformation. In *N. tabacum*, 5 out of 10^4 microspores were reporter gene positive (Stöger et al. 1995). In a similar study, the authors obtained 3.5 wheat transgenic embryos from a population of 10^6 microspores (Folling and Olesen 2001). The direct DNA transfer during microspore embryogenesis has also been employed in various crops, including rapeseed (Miki et al. 1989), barley (Olsen 1991), and tobacco (Resch et al. 2009), whereas electroporation and PEG-mediated poration has been reported in rapeseed (Jardinaud et al. 1993), barley (Vischi and Marchetti 1997), and maize (Fennell and Hauptmann 1992). Recently, Eudes and Chugh (2009) and Chugh et al. (2009) successfully transformed wheat microspores using cell-penetrating peptides with plasmid DNA and Chauhan and Khurana (2011) incorporated *HAV1* gene in wheat to obtain transgenic plants with drought-tolerant ability.

5.5   Synthetic or Artificial Seed Production

Synthetic or artificial seeds are referred as somatic embryos encapsulated in a protective coating that functionally mimic seed and have the ability to develop into normal plants under suitable in-vitro cultural conditions. The encapsulation (protective coating) is usually achieved using calcium alginate (brown algae) to protect somatic embryos during handling and from microorganism and desiccation. The whole process starts with the production of mature somatic embryos which are then immersed in a bath of calcium salts yielding encapsulated somatic embryos with clear hydrated beads. The encapsulation gel also provides required nutrients to the developing embryos just like an artificial endosperm. This technology has been successfully used to generate plants from microspore-derived embryos in wheat (Datta and Schmid 1996) and barley (Datta and Potrykus 1989) along with other monocot and dicot species.

References

Ahmad I, Day JP, MacDonald MV, Ingram DS (1991) Haploid culture and UV mutagenesis in rapid-cycling *Brassica napus* for the generation of resistance to chlorsulfuron and *Alternaria brassicicola*. Ann Bot 67(6):519–521

Alavijeh MK, Ebadi A, Omidi M (2012) Embryogenic callus production in four grapevine (*Vitis vinifera* L.) cultivars using anther and whole flower culture. Iran J Hortic Sci 43(1):1–11

Bagheri N, Babaeian-Jelodar N, Ghanbari A (2009) Evaluation of effective factors in anther culture of Iranian rice (*Oryza sativa* L.) cultivars. Biharean Biol 3(2):119–124

Barakat MN, Al-Doss AA, Elshafei AA, Moustafa KA, Ahmed EI (2012) Anther culture response in wheat (*Triticum aestivum* L.) genotypes with HMW alleles. Cereal Res Commun 40(4): 583–591. doi:10.1556/crc.40.2012.0011

Barbulescu DM, Burton WA, Salisbury PA (2011) Pluronic F-68: an answer for shoot regeneration recalcitrance in microspore-derived *Brassica napus* embryos. In Vitro Cell Dev Biol Plant 47(2):282–288. doi:10.1007/s11627-011-9353-8

Barro F, Fernandez-Escobar J, De La Vega M, Martin A (2001) Doubled haploid lines of Brassica carinata with modified erucic acid content through mutagenesis by EMS treatment of isolated microspores. Plant Breed 120(3):262–264. doi:10.1046/j.1439-0523.2001.00602.x

Basay S, Seniz V, Ellialtioglu S (2011) Obtaining dihaploid lines by using anther culture in the different eggplant cultivars. J Food Agric Environ 9(2):188–190

Belinskaya EV (2010) Genotypic features of morphogenesis in spring barley anther culture. Cytol Genet 44(2):103–107. doi:10.3103/s0095452710020052

Belmonte M, Elhiti M, Waldner B, Stasolla C (2010) Depletion of cellular brassinolide decreases embryo production and disrupts the architecture of the apical meristems in *Brassica napus* microspore-derived embryos. J Exp Bot 61(10):2779–2794. doi:10.1093/jxb/erq110

Belmonte M, Elhiti M, Ashihara H, Stasolla C (2011) Brassinolide-improved development of *Brassica napus* microspore-derived embryos is associated with increased activities of purine and pyrimidine salvage pathways. Planta 233(1):95–107. doi:10.1007/s00425-010-1287-6

Benelli C, Germana MA, Ganino T, Beghe D, Fabbri A (2010) Morphological and anatomical observations of abnormal somatic embryos from anther cultures of Citrus reticulata. Biol Plant 54(2):224–230. doi:10.1007/s10535-010-0040-0

Berber M, Yildiz M, Abak K (2012) In: Litz RE, Folta KM, Talon M, Pliego Alfaro F (eds) Effects of irradiation doses on haploid embryo and plant production in naked and shelled seed pumpkins. Acta Hortic 929:381–384

Bhat JG, Murthy HN (2008) Haploid plant regeneration from unpollinated ovule cultures of niger (*Guizotia abyssinica* (L. f.) cass.). Russ J Plant Physiol 55(2):241–245. doi:10.1134/s1021443708020118

Bhowmik P, Dirpaul J, Polowick P, Ferrie AMR (2011) A high throughput *Brassica napus* microspore culture system: influence of percoll gradient separation and bud selection on embryogenesis. Plant Cell Tiss Org 106(2):359–362. doi:10.1007/s11240-010-9913-3

Bilynska OV, Dulnyev PG (2012) Peculiarities of morphogenesis in spring barley anther culture in vitro on the nutrient media containing chemically modified starches. Fiziologiya i Biokhimia Kulturnykh Rastenii 44(5):440–448

Bjornstad A, Skinnes H, Thoresen K (1993) Comparisons between doubled haploid lines produced by anther culture, the *Hordeum bulbosum*-method and lines produced by single seed descent in barley crosses. Euphytica 66(1–2):135–144. doi:10.1007/bf00023518

Bobkov SV (2010) Isolated pea anther culture. Russ Agric Sci 36(6):413–416. doi:10.3103/s1068367410060078

Bouatrous Y, Elhady EAAA, Djekoun A, Yekhlef N (2010) Production of haploid green plants by intergeneric crossing of *Triticum durum* Desf * Zea mays L. Am Eurasian J Agric Environ Sci 7(5):512–517

Broughton S (2008) Ovary co-culture improves embryo and green plant production in anther culture of Australian spring wheat (*Triticum aestivum* L.). Plant Cell Tissue Organ Cult 95(2):185–195. doi:10.1007/s11240-008-9432-7

Broughton S (2011) The application of n-butanol improves embryo and green plant production in anther culture of Australian wheat (*Triticum aestivum* L.) genotypes. Crop Pasture Sci 62(10):813–822. doi:10.1071/cp11204

Burbulis N, Blinstrubiene A (2011) Genotypic and exogenous factors affecting linseed (*Linum usitatissimum* L.) anther culture. J Food Agric Environ 9(3–4):364–367

Burbulis N, Blinstrubiene A, Kupriene R, Zilenaite L (2009) Effect of genotype and medium composition on flax (*Linum usitatissimum* L.) anther culture. Agron Res 7(Spl. Iss. 1):204–209

Burbulis N, Blinstrubiene A, Masiene R, Jonytiene V (2012) Influence of genotype, growth regulators and sucrose concentration on linseed (*Linum usitatissimum* L.) anther culture. J Food Agric Environ 10(3–4):764–767

Castillo AM, Cistue L, Romagosa I, Valles M (2001) Low responsiveness of six-rowed genotypes to androgenesis in barley does not have a pleiotropic basis. Genome 44(5):936–940. doi:10.1139/gen-44-5-936

Chaitali S, Singh RP (2011) Anther culture response in boro rice hybrids. Asian J Biotechnol 3(5):470–477. doi:10.3923/ajbkr.2011.470.477

Chauhan H, Khurana P (2011) Use of doubled haploid technology for development of stable drought tolerant bread wheat (*Triticum aestivum* L.) transgenics. Plant Biotechnol J 9(3):408–417. doi:10.1111/j.1467-7652.2010.00561.x

Chen QF, Wang CL, Lu YM, Shen M, Afza R, Duren MV, Brunner H (2001) Anther culture in connection with induced mutations for rice improvement. Euphytica 120(3):401–408. doi:10.1023/a:1017518702176

Chen C, Liu R, Lei H, Liu S, Wang Y (2010a) Effect of low temperature pre-treatment on embryoid induction rate of tobacco anther culture. Guizhou Agric Sci 5:16–18

Chen D, Li S, Xiang Y, Zhang Z, Peng Y (2010b) Preliminary studies on the effects of plant phytosulfokine PSK-alpha in rice anther culture. Southwest China J Agric Sci 23(5):1447–1450

Chen S, Wang X, Tang H, Qian X (2011) Distant hybridization technology of wheat * maize system and evaluation of its application in haploid breeding. Mol Plant Breed 9(4):519–524

Chu CG, Xu SS, Friesen TL, Faris JD (2008) Whole genome mapping in a wheat doubled haploid population using SSRs and TRAPs and the identification of QTL for agronomic traits. Mol Breed 22(2):251–266. doi:10.1007/s11032-008-9171-9

Chugh A, Amundsen E, Eudes F (2009) Translocation of cell-penetrating peptides and delivery of their cargoes in triticale microspores. Plant Cell Rep 28(5):801–810. doi:10.1007/s00299-009-0692-4

Cistue L, Romagosa I, Batlle F, Echavarri B (2009) Improvements in the production of doubled haploids in durum wheat (*Triticum turgidum* L.) through isolated microspore culture. Plant Cell Rep 28(5):727–735. doi:10.1007/s00299-009-0690-6

Clarke HJ, Kumari M, Khan TN, Siddique KHM (2011) Poorly formed chloroplasts are barriers to successful interspecific hybridization in chickpea following in vitro embryo rescue. Plant Cell Tissue Organ Cult 106(3):465–473. doi:10.1007/s11240-011-9944-4

Corral-Martinez P, Segui-Simarro JM (2012) Efficient production of callus-derived doubled haploids through isolated microspore culture in eggplant (*Solanum melongena* L.). Euphytica 187(1):47–61. doi:10.1007/s10681-012-0715-z

Courtois B (1993) Comparison of single seed descent and anther culture-derived lines of 3 single crosses of rice. Theor Appl Genet 85(5):625–631

Croser JS, Lulsdorf MM, Grewal RK, Usher KM, Siddique KHM (2011) Isolated microspore culture of chickpea (*Cicer arietinum* L.): induction of androgenesis and cytological analysis of early haploid divisions. In Vitro Cell Dev Biol Plant 47(3):357–368. doi:10.1007/s11627-011-9346-7

Cuthbert JL, Somers DJ, Brule-Babel AL, Brown PD, Crow GH (2008) Molecular mapping of quantitative trait loci for yield and yield components in spring wheat (*Triticum aestivum* L.). Theor Appl Genet 117(4):595–608. doi:10.1007/s00122-008-0804-5

Danci M, Tudor-Radu CM, Alda S, Danci O (2011) Effect of culture medium and cultivars on callus formation and plants regeneration from anthers and immature embryos culture of wheat (*Triticum aestivum* L.). J Hortic Sci Biotechnol 15(4):110–112

Datta SK, Potrykus I (1989) Artificial seeds in barley: encapsulation of microspore-derived embryos. Theor Appl Genet 77(6):820–824. doi:10.1007/BF00268333

Datta SK, Schmid J (1996) Prospects of artificial seeds from microspore derived embryos of cereals. In: Jain SM, Sopory SK, Veilleux RE (eds) In vitro haploid production in higher plants, vol 2. Kluwer Academic, Amsterdam, pp 351–363

De Cesaro T, Baggio MI, Zanetti SA, Suzin M, Augustin L, Brammer SP, Iorczeski EJ, Milach SCK (2009) Haplodiploid androgenetic breeding in oat: genotypic variation in anther size and microspore development stage. Sci Agric 66(1):118–122

de Moraes AP, Bered F, de Carvalho FIF, Kaltchuk-Santos E (2008) Morphological markers for microspore developmental stage in maize. Braz Arch Biol Technol 51(5):911–916

Diao W-P, Jia Y-Y, Song H, Zhang X-Q, Lou Q-F, Chen J-F (2009) Efficient embryo induction in cucumber ovary culture and homozygous identification of the regenetants using SSR markers. Sci Hortic 119(3):246–251. doi:10.1016/j.scienta.2008.08.016

Doi H, Yokoi S, Hikage T, Nishihara M, Tsutsumi K-i, Takahata Y (2011) Gynogenesis in gentians (*Gentiana triflora*, *G. scabra*): production of haploids and doubled haploids. Plant Cell Rep 30(6):1099–1106. doi:10.1007/s00299-011-1017-y

Dormann P, Hoffmannbenning S, Balbo I, Benning C (1995) Isolation and characterization of an arabidopsis mutant deficient in the thylakoid lipid digalactosyl diacylglycerol. Plant Cell 7(11):1801–1810. doi:10.2307/3870188

Dubas E, Wedzony M, Petrovska B, Salaj J, Zur I (2010) Cell structural reorganization during induction of androgenesis in isolated microspore cultures of triticale (x triticosecale wittm.). Acta Biol Cracov Bot 52(1):73–86. doi:10.2478/v10182-010-0010-z

Dunwell JM (2010) Haploids in flowering plants: origins and exploitation. Plant Biotechnol J 8(4):377–424. doi:10.1111/j.1467-7652.2009.00498.x

Dyulgerova B, Valcheva D, Dimova D (2010) Anther culture response in winter barley (*Hordeum vulgare* L.). Genet Breed 39(1–2):45–49

Eder J, Chalyk S (2002) In vivo haploid induction in maize. Theor Appl Genet 104(4):703–708. doi:10.1007/s00122-001-0773-4

El-Hennawy MA, Abdalla AF, Shafey SA, Al-Ashkar IM (2011) Production of doubled haploid wheat lines (*Triticum aestivum* L.) using anther culture technique. Ann Agric Sci 56(2):69–77

Ellaltoglu S, Basay S, Kusvuran S (2012) Investigations on the pollen dimorphism and its relationship with anther culture in eggplant (Patlcanda Polen dimorfizmi ve anter kulturu iliskisinin incelenmesi.). TABAD Tarm Bilimleri Arastrma Dergisi 5(1):149–152

Ercan N, Sensoy FA (2012) In: Geelen D (ed) Determination of the optimum microspore development stage and optimum culture medium in asparagus (*Asparagus officinalis* var. altilis L.) for anther culture. Acta Hortic 961:153–158

Esteves P, Marchand S, Sangare M, Belzile F (2010) Effects of pretreatments and hormone regime on green plant production in barley isolated microspore culture. In Vitro Cell Dev Biol Anim 46:S112

Eudes F, Chugh A (2009) An overview of triticale doubled haploids. Adv Haploid Prod Higher Plants. doi:10.1007/978-1-4020-8854-4_6

Faostat (2010) http://faostat.fao.org

Fennell A, Hauptmann R (1992) Electroporation and PEG delivery of DNA into maize microspores. Plant Cell Rep 11(11):567–570. doi:10.1007/BF00233094

Folling L, Olesen A (2001) Transformation of wheat (*Triticum aestivum* L.) microspore-derived callus and microspores by particle bombardment. Plant Cell Rep 20(7):629–636. doi:10.1007/s002990100371

Forodi BR, Hassandokht M, Kashi A, Sepahvand N (2009) Influence of spermidine on haploid plant production in Iranian onion (*Allium cepa* L.) populations through in vitro culture. Hortic Environ Biotechnol 50(5):461–466

Froelicher Y, Bassene J-B, Jedidi-Neji E, Dambier D, Morillon R, Bernardini G, Costantino G, Ollitrault P (2007) Induced parthenogenesis in mandarin for haploid production: induction procedures and genetic analysis of plantlets. Plant Cell Rep 26(7):937–944. doi:10.1007/s00299-007-0314-y

Geng X, Wei Z, Yao X, Zhao Y (2010) Anther culture of *Medicago sativa* L. and ploidy detection. Acta Agric Sin 18(5):714–718

Germana MA (2011) Gametic embryogenesis and haploid technology as valuable support to plant breeding. Plant Cell Rep 30(5):839–857. doi:10.1007/s00299-011-1061-7

Ghazanfari P, Abdollahi MR, Moieni A, Moosavi SS (2012) Effect of plant-derived smoke extract on in vitro plantlet regeneration from rapeseed (*Brassica napus* L. cv. Topas) microspore-derived embryos. Int J Plant Prod 6(3):309–324

Godbole M, Murthy HN (2012) Parthenogenetic haploid plants using gamma irradiated pollen in snapmelon (*Cucumis melo* var. momordica). Plant Cell Tissue Organ Cult 109(1):167–170. doi:10.1007/s11240-011-0066-9

Gorecka K, Krzyzanowska D, Kiszczak W, Kowalska U (2009) Plant regeneration from carrot (*Daucus carota* L.) anther culture derived embryos. Acta Physiol Plant 31(6):1139–1145. doi:10.1007/s11738-009-0332-1

Gorecka K, Kowalska U, Krzyzanowska D, Kiszczak W (2010) Obtaining carrot (*Daucus carota* L.) plants in isolated microspore cultures. J Appl Genet 51(2):141–147. doi:10.1007/bf03195722

Grouh MSH, Vandati K, Lotfi M, Hassani D, Biranvand NP (2011) Production of haploids in persian walnut through parthenogenesis induced by gamma-irradiated pollen. J Am Soc Hortic Sci 136(3):198–204

Grozeva S, Rodeva V, Todorova V, Pandeva R (2009) Obtaining of pepper plants via anther culture. Genet Breed 38(3–4):25–31

Gu J, Liu K, Li S, Tian Y, Yang H, Yang M (2008) Study on the in vitro culture of cut plants in wheat haploid embryo induction by a wheat * maize cross. Front Agric China 2(4):391–395. doi:10.1007/s11703-008-0070-y

Guerrand J, Euzen M, Courand D, Bodin M, Tanguy JL, Menard V (2012) Microscopic study of artichoke (*Cynara scolymus* L.) floral pieces in order to develop a haploid production technique. In: Bazinet C, Mabeau S (eds) VII International symposium on artichoke, cardoon and their wild relatives. Acta Hortic 942:165–169

Hassawi DS, Abu-Mallouh SA, Al-Abbadi AA, Shatnawi MA (2012) Organelles genome stability of wheat plantlets produced by anther culture. Afr J Biotechnol 11(22):6018–6026

He P, Li JZ, Zheng XW, Shen LS, Lu CF, Chen Y, Zhu LH (2001) Comparison of molecular linkage maps and agronomic trait loci between DH and RIL populations derived from the same rice cross. Crop Sci 41(4):1240–1246. doi:10.2135/cropsci2001.4141240x

Heidari B, Saeidi G, Tabatabaei BES, Suenaga K (2012) QTLs involved in plant height, peduncle length and heading date of wheat (*Triticum aestivum* L.). J Agric Sci Technol 14(5): 1093–1104

Herath HMI, Bandara DC (2011) Anther culture performance in selected high yielding indica (of Sri Lanka) and japonica rice varieties and their inter sub-specific hybrids. J Natl Sci Found Sri Lanka 39(2):149–154

Houben A, Sanei M, Pickering R (2011) Barley doubled-haploid production by uniparental chromosome elimination. Plant Cell Tissue Organ Cult 104(3):321–327. doi:10.1007/s11240-010-9856-8

Huang B (1992) Genetic manipulation of microspores and microspore-derived embryos. In Vitro Cell Dev Biol Plant 28(2):53–58

Hyde PT, Earle ED, Mutschler MA (2012) Doubled haploid onion (*Allium cepa* L.) lines and their impact on hybrid performance. Hortic Sci 47(12):1690–1695

Irikova T, Grozeva S, Popov P, Rodeva V, Todorovska E (2011a) In vitro response of pepper anther culture (*Capsicum annuum* L.) depending on genotype, nutrient medium and duration of cultivation. Biotechnol Biotechnol Equip 25(4):2604–2609. doi:10.5504/bbeq.2011.0090

Irikova T, Grozeva S, Rodeva V (2011b) Anther culture in pepper (*Capsicum annuum* L.) in vitro. Acta Physiol Plant 33(5):1559–1570. doi:10.1007/s11738-011-0736-6

Islam SMS (2010a) The effect of colchicine pretreatment on isolated microspore culture of wheat (*Triticum aestivum* L.). Aust J Crop Sci 4(9):660–665

Islam SMS (2010b) Effect of embryoids age, size and shape for improvement of regeneration efficiency from microspore-derived embryos in wheat (*Triticum aestivum* L.). Plant Omics 3(5):149–153

Islam SMS (2010c) The role of drought stress on anther culture of wheat (*Triticum aestivum* L.). Plant Tissue Cult Biotechnol 20(1):55–61

Jacquard C, Mazeyrat-Gourbeyre F, Devaux P, Boutilier K, Baillieul F, Clement C (2009a) Microspore embryogenesis in barley: anther pre-treatment stimulates plant defence gene expression. Planta 229(2):393–402. doi:10.1007/s00425-008-0838-6

Jacquard C, Nolin F, Hecart C, Grauda D, Rashal I, Dhondt-Cordelier S, Sangwan RS, Devaux P, Mazeyrat-Gourbeyre F, Clement C (2009b) Microspore embryogenesis and programmed cell death in barley: effects of copper on albinism in recalcitrant cultivars. Plant Cell Rep 28(9):1329–1339. doi:10.1007/s00299-009-0733-z

Jaeger K, Bartok T, Oerdoeg V, Barnabas B (2010) Improvement of maize (*Zea mays* L.) anther culture responses by algae-derived natural substances. S Afr J Bot 76(3):511–516. doi:10.1016/j.sajb.2010.03.009

Jardinaud M-F, Souvré A, Alibert G (1993) Transient GUS gene expression in *Brassica napus* electroporated microspores. Plant Sci 93(1–2):177–184. doi:10.1016/0168-9452(93)90047-4

Jo MH, Ham IK, Park MY, Kim TI, Lim YP, Lee EM (2012) Seed production ability of doubled haploid plants through microspore culture in Chinese cabbage (*Brassica rapa* L. ssp pekinensis) introduced from China. Korean J Hortic Sci 30(5):573–578. doi:10.7235/hort.2012.12004

Kantartzi SK, Roupakias DG (2009) In vitro gynogenesis in cotton (Gossypium sp.). Plant Cell Tissue Organ Cult 96(1):53–57. doi:10.1007/s11240-008-9459-9

Kasha KJ, Simion E, Oro R, Yao QA, Hu TC, Carlson AR (2001) An improved in vitro technique for isolated microspore culture of barley. Euphytica 120(3):379–385. doi:10.1023/a:1017564100823

Kelm C, Ghaffary SMT, Bruelheide H, Roder MS, Miersch S, Weber WE, Kema GHJ, Saal B (2012) The genetic architecture of seedling resistance to Septoria tritici blotch in the winter wheat doubled-haploid population Solitar x Mazurka. Mol Breed 29(3):813–830. doi:10.1007/s11032-011-9592-8

Khan MA, Ahmad J (2011) In vitro wheat haploid embryo production by wheat x maize cross system under different environmental conditions. Pak J Agric Sci 48(1):49–53

Kielkowska A, Adamus A (2010) In vitro culture of unfertilized ovules in carrot (*Daucus carota* L.). Plant Cell Tissue Organ Cult 102(3):309–319. doi:10.1007/s11240-010-9735-3

Kim J, Lee S-S (2012) Identification of monogenic dominant male sterility and its suppressor gene from an induced mutation using a Broccoli (*Brassica oleracea* var. italica) microspore culture. Hortic Environ Biotechnol 53(3):237–241. doi:10.1007/s13580-012-0091-6

Kim M, Park EJ, An D, Lee Y (2013) High-quality embryo production and plant regeneration using a two-step culture system in isolated microspore cultures of hot pepper (*Capsicum annuum* L.). Plant Cell Tissue Organ Cult 112(2):191–201. doi:10.1007/s11240-012-0222-x

Kiszczak W, Krzyzanowska D, Strycharczuk K, Kowalska U, Wolko B, Gorecka K (2011) Determination of ploidy and homozygosity of carrot plants obtained from anther cultures. Acta Physiol Plant 33(2):401–407. doi:10.1007/s11738-010-0559-x

Kiviharju EM (2009) Anther culture derived doubled haploids in oat. In: Touraev A, Forster BP, Jain SM (eds) Advances in haploid production in higher plants. Springer, Dordrecht, pp 171–178. doi:10.1007/978-1-4020-8854-4_14

Koleva-Gudeva L, Trajkova F, Dimeska G, Spasenoski M (2009) Androgenesis efficiency in anther culture of pepper (*Capsicum annuum* L.). In: Krasteva L, Panayotov N (eds) IV Balkan symposium on vegetables and potatoes. Acta Hortic 830:183–190

Kott LS (1996) Production of mutants using the rapeseed doubled haploid system. In: Induced mutation and molecular techniques for crop improvement. IAEA/FAO Proceedings of an international symposium on the use of induced mutations and molecular techniques for crop improvement, Vienne, Austria, pp 505–515

Kour A, Bhatt U, Grewal S, Singh NK, Khanna VK (2008) Effect of wheat and maize genotypes for wheat haploid production in wheat x maize crosses. Indian J Genet Plant Breed 68(2):201–203

Krystyna G, Dorota K, Urszula K, Waldemar K (2010) The technology of obtaining of carrot homozygous plants using the anther cultures. In Vitro Cell Dev Biol Anim 46:S189

Krzewska M, Czyczylo-Mysza I, Dubas E, Golebiowska-Pikania G, Golemiec E, Stojalowski S, Chrupek M, Zur I (2012) Quantitative trait loci associated with androgenic responsiveness in triticale (xTriticosecale Wittm.) anther culture. Plant Cell Rep 31(11):2099–2108. doi:10.1007/s00299-012-1320-2

Kuchel H, Williams K, Langridge P, Eagles HA, Jefferies SP (2007a) Genetic dissection of grain yield in bread wheat. II. QTL-by-environment interaction. Theor Appl Genet 115(7):1015–1027. doi:10.1007/s00122-007-0628-8

Kuchel H, Williams KJ, Langridge P, Eagles HA, Jefferies SP (2007b) Genetic dissection of grain yield in bread wheat. I. QTL analysis. Theor Appl Genet 115(8):1029–1041. doi:10.1007/s00122-007-0629-7

Lantos C, Juhasz AG, Somogyi G, Otvos K, Vagi P, Mihaly R, Kristof Z, Somogyi N, Pauk J (2009) Improvement of isolated microspore culture of pepper (*Capsicum annuum* L.) via co-culture with ovary tissues of pepper or wheat. Plant Cell Tissue Organ Cult 97(3):285–293. doi:10.1007/s11240-009-9527-9

Lantos C, Juhasz AG, Vagi P, Mihaly R, Kristof Z, Pauk J (2012) Androgenesis induction in microspore culture of sweet pepper (*Capsicum annuum* L.). Plant Biotechnol Rep 6(2):123–132. doi:10.1007/s11816-011-0205-0

Lee S-S, Lee S-A, Yang J, Kim J (2011) Developing stable progenies of xBrassicoraphanus, an intergeneric allopolyploid between *Brassica rapa* and *Raphanus sativus*, through induced mutation using microspore culture. Theor Appl Genet 122(5):885–891. doi:10.1007/s00122-010-1494-3

Li S, Huang Y, Xiao Y, Zhang B (2012) An observe of "Anther Culture in Pepper (*Capsicum annuum* L.) in vitro". China Vegetables (20):1–6

Liu Y, Li L, Chen S (2009) Genetic analysis of double haploid population from maize bio-induced parthenogenesis. J China Agric Univ 14(1):56–60

Liu B, Miao J, Wang W, Zhang Z, Yang Y, Zhang Y, Wu X (2010) Research progress in onion (*Allium cepa* L.) haploid culture. China Vegetables (6):8–13

Ljevnaic B, Kondic-Spika A, Kobiljski B, Hristov N (2009) Cytological characteristics of regenerants obtained from anther culture of wheat (*Triticum aestivum* L.). In: Mihailovic D, Miloradov MV (eds) Environmental, health and humanity issues in the down Danubian region: multidisciplinary approaches. World Scientific Publishing, Singapore, doi:10.1142/9789812834409_0015

Lotfi M, Salehi S (2008) Detection of cucumber parthenogenic haploid embryos by floating the immature seeds in liquid medium. In: Cucurbitaceae 2008: Proceedings of the IXth Eucarpia meeting on genetics and breeding of Cucurbitaceae

Ma J, Zhang B (2008) Effects of glutamine, proline, and exogenous hormones on the anther culture of alfalfa. Acta Agric Sin 16(4):364–369

Ma ZQ, Steffenson BJ, Prom LK, Lapitan NLV (2000) Mapping of quantitative trait loci for Fusarium head blight resistance in barley. Phytopathology 90(10):1079–1088. doi:10.1094/phyto.2000.90.10.1079

Malik MR, Krochko JE (2009) Gene expression profiling of microspore embryogenesis in *Brassica napus*. Adv Haploid Prod Higher Plants. doi:10.1007/978-1-4020-8854-4_8

Maluszynski M, Ahloowalia BS (2000) Mutation techniques in plant breeding. In: Moriaty M, Edington M, Mothersill C, Ward JF, Seymour C, Fry RJM (eds) Radiation research, vol 2. Cong Proceed, pp 251–254

Mankowska G, Rutkowska-Krause I, Luwanska A, Wielgus K, Makowiecka J (2011) Use of anther culture to improve flax (*Linum usitatissimum* L.) resistance to Fusarium wilt (Wykorzystanie kultury pylnikowej w celu polepszenia odpornosci na fuzarioze lnu woknistego (Linum ustatissimum L.)). Biuletyn Instytutu Hodowli i Aklimatyzacji Roslin (260/261):333–339

Marcinska I, Nowakowska A, Skrzypek E, Czyczylo-Mysza I (2013) Production of double haploids in oat (*Avena sativa* L.) by pollination with maize (*Zea mays* L.). Cent Eur J Biol 8(3):306–313. doi:10.2478/s11535-013-0132-2

Miki B, Huang B, Bird S, Kemble R, Simmonds D, Keller W (1989) A procedure for the microinjection of plant cells and protoplasts. J Tissue Cult Methods 12(4):139–144. doi:10.1007/BF01404440

Mohammadi PP, Moieni A, Ebrahimi A, Javidfar F (2012) Doubled haploid plants following colchicine treatment of microspore-derived embryos of oilseed rape (*Brassica napus* L.). Plant Cell Tiss Org 108(2):251–256. doi:10.1007/s11240-011-0036-2

Motallebi-Azar A (2010) Androgenic response of tomato (*Lycopersicon esculentum* Mill.) lines and their hybrids to anther culture. Russ Agric Sci 36(4):250–258

Motallebi-Azar A, Panahandeh J (2010) Effects of colchicine and cold duration pretreatments on androgenesis responses of tomato (*Lycopersicon esculentum* Mill) via anther culture. Russ Agric Sci 36(5):338–341. doi:10.3103/s106836741005006x

Mozgova GV, Zaitseva OI, Lemesh VA (2012) Structural changes in chloroplast genome accompanying albinism in anther culture of wheat and triticale. Cereal Res Commun 40(4):467–475. doi:10.1556/crc.40.2012.0007

Na H, Park S, Hwang G, Yoon M-K, Chun C (2009) Medium, AgNO$_3$, activated charcoal and NAA effects on microspore culture in *Brassica rapa*. Korean J Hortic Sci 27(4):657–661

Na H, Hwang G, Kwak J-H, Yoon MK, Chun C (2011a) Microspore derived embryo formation and doubled haploid plant production in broccoli (*Brassica oleracea* L. var italica) according to nutritional and environmental conditions. Afr J Biotechnol 10(59):12535–12541. doi:10.5897/ajb10.1287

Na H, Kwak J-H, Chun C (2011b) The effects of plant growth regulators, activated charcoal, and AgNO$_3$ on microspore derived embryo formation in broccoli (*Brassica oleracea* L. var. italica). Hortic Environ Biotechnol 52(5):524–529. doi:10.1007/s13580-011-0034-7

Nelson MN, Mason AS, Castello M-C, Thomson L, Yan G, Cowling WA (2009a) Microspore culture preferentially selects unreduced (2n) gametes from an interspecific hybrid of *Brassica napus* L. x Brassica carinata. Theor Appl Genet 119(5):953–953. doi:10.1007/s00122-009-1126-y

Nelson MN, Mason AS, Castello M-C, Thomson L, Yan G, Cowling WA (2009b) Microspore culture preferentially selects unreduced (2n) gametes from an interspecific hybrid of *Brassica napus* L. x *Brassica carinata* Braun. Theor Appl Genet 119(3):497–505. doi:10.1007/s00122-009-1056-8

Obert B, Uvackova L, Pret'ova A (2009) Maize doubled haploids via anther and microspore culture. In: Danforth AT (ed) Corn crop production: growth, fertilization and yield. Nova Science, Hauppauge, NY, pp 333–343

Ochoa-Alejo N (2012) Anther culture of chili pepper (*Capsicum* spp.). Methods Mol Biol 877:227–231

Olsen FL (1991) Isolation and cultivation of embryogenic microspores from barley (*Hordeum vulgare* L). Hereditas 115(3):255–266

Olszewska D, Kisiala A, Nowaczyk P (2011) The assessment of doubled haploid lines obtained in pepper (*Capsicum annuum* L.) anther culture. Folia Hortic 23(2):93–99

Pathirana R, Frew T, Hedderley D, Timmerman-Vaughan G, Morgan E (2011) Haploid and doubled haploid plants from developing male and female gametes of *Gentiana triflora*. Plant Cell Rep 30(6):1055–1065. doi:10.1007/s00299-011-1012-3

Ponitka A, Slusarkiewicz-Jarzina A (2011) Production of spontaneous and induced doubled-haploid lines of winter triticale obtained through anther culture (Otrzymywanie spontanic-znych i indukowanych linii podwojonych haploidow pszenzyta ozimego z wykorzystaniem kultur pylnikowych.). Biuletyn Instytutu Hodowli i Aklimatyzacji Roslin (260/261):183–191

Poon S, Heath RL, Clarke AE (2012) A chimeric arabinogalactan protein promotes somatic embryogenesis in cotton cell culture. Plant Physiol 160(2):684–695. doi:10.1104/pp.112.203075

Prem D, Solis M-T, Barany I, Rodriguez-Sanz H, Risueno MC, Testillano PS (2012) A new microspore embryogenesis system under low temperature which mimics zygotic embryogenesis initials, expresses auxin and efficiently regenerates doubled-haploid plants in *Brassica napus*. BMC Plant Biol 12:127. doi:10.1186/1471-2229-12-127

Prodanovic S, Matzk F, Zoric D (2008) Effect of dicamba on wheat haploid embryo development. Cereal Res Commun 36(1):43–51. doi:10.1556/crc36.2008.1.5

Rakha MT, Metwally EI, Moustafa SA, Etman AA, Dewir YH (2012) Evaluation of regenerated strains from six Cucurbita interspecific hybrids obtained through anther and ovule in vitro cultures. Aust J Crop Sci 6(1):23–30

Redha A, Suleman P (2011) Effects of exogenous application of polyamines on wheat anther cultures. Plant Cell Tissue Organ Cult 105(3):345–353. doi:10.1007/s11240-010-9873-7

Redha A, Talaat A (2008) Improvement of green plant regeneration by manipulation of anther culture induction medium of hexaploid wheat. Plant Cell Tissue Organ Cult 92(2):141–146. doi:10.1007/s11240-007-9315-3

Ren JP, Wang XG, Yin J (2010) Dicamba and sugar effects on callus induction and plant regeneration from mature embryo culture of wheat. Agr Sci China 9(1):31–37. doi:10.1016/s1671-2927(09)60064-x

Resch T, Ankele E, Badur R, Reiss B, Herberle-Bors E, Touraev A (2009) Immature pollen as a target for gene targeting. In: Touraev A, Forster B, Jain SM (eds) Advances in haploid production in higher plants. Springer, Dordrecht, pp 307–317. doi:10.1007/978-1-4020-8854-4_25

Rodriguez-Serrano M, Barany I, Prem D, Coronado MJ, Risueno MC, Testillano PS (2012) NO, ROS, and cell death associated with caspase-like activity increase in stress-induced microspore embryogenesis of barley. J Exp Bot 63(5):2007–2024. doi:10.1093/jxb/err400

Rosenbaumova R, Krahulcova A, Krahulec F (2012) The intriguing complexity of parthenogenesis inheritance in *Pilosella rubra* (Asteraceae, Lactuceae). Sex Plant Reprod 25(3):185–196. doi:10.1007/s00497-012-0190-7

Salas P, Prohens J, Segui-Simarro JM (2011) Evaluation of androgenic competence through anther culture in common eggplant and related species. Euphytica 182(2):261–274. doi:10.1007/s10681-011-0490-2

Salas P, Rivas-Sendra A, Prohens J, Segui-Simarro JM (2012) Influence of the stage for anther excision and heterostyly in embryogenesis induction from eggplant anther cultures. Euphytica 184(2):235–250. doi:10.1007/s10681-011-0569-9

Santra M, Ankrah N, Santra DK, Kidwell KK (2012) An improved wheat microspore culture technique for the production of doubled haploid plants. Crop Sci 52(5):2314–2320. doi:10.2135/cropsci2012.03.0141

Sayem MA, Maniruzzaman M, Siddique SS, Al-Amin M (2010) In vitro shoot regeneration through anther culture of Brassica spp. Bangladesh J Agric Res 35(2):331–341. doi:10.3329/bjar.v35i2.5896

Segui-Simarro JM, Corral-Martinez P, Corredor E, Raska I, Testillano PS, Risueno MC (2011) A change of developmental program induces the remodeling of the interchromatin domain during microspore embryogenesis in *Brassica napus* L. J Plant Physiol 168(8):746–757. doi:10.1016/j.jplph.2010.10.014

Shabana M, Mari SN, Mari AK, Gaddi NH (2010) Induction of callus through anther and ovule culture in upland cotton (*Gossypium hirsutum* L.). World Appl Sci J 8(Spl Issue):76–79

Shere UB, Dhage SJ (2009) Studies on anther culture in tomato (*Lycopersicon esculentum*). Int J Plant Sci (Muzaffarnagar) 4(2):433–435

Shirdelmoghanloo H, Moieni A, Mousavi A (2009) Effects of embryo induction media and pretreatments in isolated microspore culture of hexaploid wheat (*Triticum aestivum* L. cv. Falat). Afr J Biotechnol 8(22):6134–6140

Sidhu PK, Davies PA (2009) Regeneration of fertile green plants from oat isolated microspore culture. Plant Cell Rep 28(4):571–577. doi:10.1007/s00299-009-0684-4

Smykalova I, Horacek J, Kubosiova M, Smirous P, Soukup A, Gasmanova N, Griga M (2012) Induction conditions for somatic and microspore-derived structures and detection of haploid status by isozyme analysis in anther culture of caraway (*Carum carvi* L.). In Vitro Cell Dev Biol Plant 48(1):30–39. doi:10.1007/s11627-011-9386-z

Song Y, Zhou S, Du X, Hua S, Wei P, Chen Y (2012) Study on the factors affecting the anther culture efficiency of wheat. J Northwest A&F Univ Nat Sci Ed 40(5):62–68

Soriano M, Cistue L, Castillo AM (2008) Enhanced induction of microspore embryogenesis after n-butanol treatment in wheat (*Triticum aestivum* L.) anther culture. Plant Cell Rep 27(5):805–811. doi:10.1007/s00299-007-0500-y

Sourdille P, Snape JW, Cadalen T, Charmet G, Nakata N, Bernard S, Bernard M (2000) Detection of QTLs for heading time and photoperiod response in wheat using a doubled-haploid population. Genome 43(3):487–494. doi:10.1139/gen-43-3-487

Stöger E, Fink C, Pfosser M, Heberle-Bors E (1995) Plant transformation by particle bombardment of embryogenic pollen. Plant Cell Rep 14(5):273–278. doi:10.1007/BF00232027

Sun N, Shi P, Wei L, Cao Y, Huang R, Li S, Liang W (2009) Genetic control analysis of triticale anther culture response. J Triticeae Crops 29(3):374–379

Supena EDJ, Custers JBM (2011) Refinement of shed-microspore culture protocol to increase normal embryos production in hot pepper (*Capsicum annuum* L.). Sci Hortic 130(4):769–774. doi:10.1016/j.scienta.2011.08.037

Suprunova T, Shmykova N (2008) In vitro induction of haploid plants in unpollinated ovules, anther and microspore culture of *Cucumis sativus*. In: Pitrat M (ed) Cucurbitaceae 2008: Proceedings of the sixth Eucarpia meeting on genetics and breeding of Cucurbitaceae. INRA, Avignon, pp 371–374

Suzuki T, Sato M, Takeuchi T (2012) Evaluation of the effects of five QTL regions on Fusarium head blight resistance and agronomic traits in spring wheat (*Triticum aestivum* L.). Breed Sci 62(1):11–17. doi:10.1270/jsbbs.62.11

Swanson EB, Herrgesell MJ, Arnoldo M, Sippell DW, Wong RSC (1989) Microspore mutagenesis and selection - canola plants with field tolerance to the imidazolinones. Theor Appl Genet 78(4):525–530. doi:10.1007/bf00290837

Szafranska K, Cvikrova M, Kowalska U, Gorecka K, Gorecki R, Martincova O, Janas KM (2011) Influence of copper ions on growth, lipid peroxidation, and proline and polyamines content in carrot rosettes obtained from anther culture. Acta Physiol Plant 33(3):851–859. doi:10.1007/s11738-010-0610-y

Takahashi Y, Yokoi S, Takahata Y (2011) Improvement of microspore culture method for multiple samples in Brassica. Breed Sci 61(1):96–98. doi:10.1270/jsbbs.61.96

Takahashi Y, Yokoi S, Takahata Y (2012) Effects of genotypes and culture conditions on micro-spore embryogenesis and plant regeneration in several subspecies of *Brassica rapa* L. Plant Biotechnol Rep 6(4):297–304. doi:10.1007/s11816-012-0224-5

Takahira J, Cousin A, Nelson MN, Cowling WA (2011) Improvement in efficiency of microspore culture to produce doubled haploid canola (*Brassica napus* L.) by flow cytometry. Plant Cell Tiss Org 104(1):51–59. doi:10.1007/s11240-010-9803-8

Tang Y, Li H, Liu J, Liu B (2009) Callus formation from anther culture in balsam pear (*Momordica charantia* L.). Am Eurasian J Agric Environ Sci 6(3):308–312

Tang Y, Li X, Li J, Ma C, Lai J, Li H (2012) Effect of different pretreatment on callus formation from anther in balsam pear (*Momordica charantia* L.). J Med Plant Res 6(17):3393–3395. doi:10.5897/jmpr11.1752

Targonska M, Hromada-Judycka A, Bolibok-Bragoszewska H, Rakoczy-Trojanowska M (2013) The specificity and genetic background of the rye (*Secale cereale* L.) tissue culture response. Plant Cell Rep 32(1):1–9. doi:10.1007/s00299-012-1342-9

Taskin H, Buyukalaca S, Keles D, Ekbic E (2011) Induction of microspore-derived embryos by anther culture in selected pepper genotypes. Afr J Biotechnol 10(75):17116–17121. doi:10.5897/ajb11.2023

Testillano PS, Coronado MJ, Thierry AM, Matthys-Rochon E, Risueno MC (2010) In situ detec-tion of Esr proteins secretion during maize microspore embryogenesis and their secretion blockage show effects on the culture progression. Funct Plant Biol 37(10):985–994. doi:10.1071/fp10066

Tomaszewska-Sowa M (2010) Cytometric analyses of sugar beet (*Beta vulgaris* L.) plants regener-ated from unfertilized ovules cultured in vitro. Electron J Pol Agric Univ 13(4):09-art 09

Touraev A, Pfosser M, Heberle-Bors E (2001) The microspore: a haploid multipurpose cell. Adv Bot Res 35:53–109. doi:10.1016/s0065-2296(01)35004-8

Tyrnov VS, Smolkina YV (2011) The new technology of maize breeding by parthenogenesis. Maize Genet Coop News Lett 85:36

Usha B, Khanna VK (2010) Pollen tube study in relation to haploid production in wheat maize crosses. Pantnagar J Res 8(1):39–42

Vischi M, Marchetti S (1997) Strong extracellular nuclease activity displayed by barley (*Hordeum vulgare* L.) uninucleate microspores. Theor Appl Genet 95(1–2):185–190. doi:10.1007/s001220050546

Wan GL, Naeem MS, Geng XX, Xu L, Li B, Jilani G, Zhou WJ (2011) Optimization of microspore embryogenesis and plant regeneration protocols for *Brassica napus*. Int J Agric Biol 13(1):83–88

Wang S, Liu Y, Khalid H, Wang Y, Shi Z, Lin F (2010) Genetic study on colchicine induction of parthenogenesis in maize. World Appl Sci J 10(2):143–146

Wang Y, Tong Y, Li Y, Zhang Y, Zhang J, Feng J, Feng H (2011) High frequency plant regeneration from microspore-derived embryos of ornamental kale (*Brassica oleracea* L. var. acephala). Sci Hortic 130(1):296–302. doi:10.1016/j.scienta.2011.06.029

Wen J, Zeng X-h, Pu Y-y, Qi L-p, Li Z-y, Tu J-x, Ma C-z, Shen J-x, Fu T-d (2010) Meiotic nondis-junction in resynthesized *Brassica napus* and generation of aneuploids through microspore culture and their characterization. Euphytica 173(1):99–111. doi:10.1007/s10681-010-0129-8

Weber BN, Hamernik AJ, Jansky SH (2012) Hybridization barriers between diploid Solanum tuberosum and wild Solanum raphanifolium. Genet Resour Crop Evol 59(7):1287–1293. doi:10.1007/s10722-012-9883-x

Wei J, Chen M, Liu Z, Zhu L (2010) Identification method of maize haploid induced by partheno-genesis inducer. Chin Seed 1(1):40–42

Weyen J, Orsini J, Gnad H (2012) A new wheat anther culture technology preventing albinism in the EU winter wheat gene pool. In Vitro Cell Dev Biol Anim 48:74

Winarto B, da Silva JAT (2011) Microspore culture protocol for Indonesian *Brassica oleracea*. Plant Cell Tiss Org 107(2):305–315. doi:10.1007/s11240-011-9981-z

Yadollahi A, Abdollahi MR, Moieni A, Danaee M (2011) Effects of carbon source, polyethylene glycol and abscisic acid on secondary embryo induction and maturation in rapeseed *(Brassica napus* L.) microspore-derived embryos. Acta Physiol Plant 33(5):1905–1912. doi:10.1007/s11738-011-0738-4

Yahata M, Yasuda K, Nagasawa K, Harusaki S, Komatsu H, Kunitake H (2010) Production of haploid plant of 'Banpeiyu' Pummelo Citrus maxima (Burm.) Merr. by pollination with soft X-ray-irradiated pollen. J Jpn Soc Hortic Sci 79(3):239–245

Yamagishi M, Otani M, Higashi M, Fukuta Y, Fukui K, Yano M, Shimada T (1998) Chromosomal regions controlling anther culturability in rice *(Oryza sativa* L.). Euphytica 103(2):227–234. doi:10.1023/a:1018328708322

Yamashita K-i, Nakazawa Y, Namai K, Amagai M, Tsukazaki H, Wako T, Kojima A (2012) Modes of inheritance of two apomixis components, diplospory and parthenogenesis, in Chinese chive *(Allium ramosum)* revealed by analysis of the segregating population generated by back-crossing between amphimictic and apomictic diploids. Breed Sci 62(2):160–169. doi:10.1270/jsbbs.62.160

Yang Y, Yu C, Song H, Xu H, Xu Z (2011) Optimization of culture conditions of F 1 anther culture from different panicle shape japonica/indica hybrid rice. J Shenyang Agric Univ 42(3):354–357

Yuan L, Song D, Gao G, He Y (2011) Improvement of resistance to rice blast in PGMS line Y58S by molecular marker-assisted selection and anther culture. Genomics Appl Biol 30(5):620–625

Yuan SX, Su YB, Liu YM, Fang ZY, Yang LM, Zhuang M, Zhang YY, Sun PT (2012) Effects of pH, MES, arabinogalactan-proteins on microspore cultures in white cabbage. Plant Cell Tissue Organ Cult 110(1):69–76. doi:10.1007/s11240-012-0131-z

Zeng X, Wen J, Wan Z, Yi B, Shen J, Ma C, Tu J, Fu T (2010) Effects of Bleomycin on microspore embryogenesis in *Brassica napus* and detection of somaclonal variation using AFLP molecular markers. Plant Cell Tiss Org 101(1):23–29. doi:10.1007/s11240-009-9658-z

Zhang L, Zhang L, Luo J, Chen W, Hao M, Liu B, Yan Z, Zhang B, Zhang H, Zheng Y, Liu D, Yen Y (2011) Synthesizing double haploid hexaploid wheat populations based on a spontaneous alloploidization process. J Genet Genomics 38(2):89–94. doi:10.1016/j.jcg.2011.01.004

Zhao J, Zhou X, Zhang Z, Yang B, Zhou S (2010) Effects of culture media on anther culture of chili pepper *(Capsicum annuum* L.). J Hunan Agric Univ 36(2):181–187

Zhao L, Liu L, Guo H, Gu J, Zhao S, Li J, XieYongDun (2012) Combining ability analysis of anther culture traits of three wheat genotypes with high regeneration ability. J Triticeae Crops 32(3):427–430

Zur I, Dubas E, Golemiec E, Szechynska-Hebda M, Golebiowska G, Wedzony M (2009) Stress-related variation in antioxidative enzymes activity and cell metabolism efficiency associated with embryogenesis induction in isolated microspore culture of triticale (x Triticosecale Wittm.). Plant Cell Rep 28(8):1279–1287. doi:10.1007/s00299-009-0730-2

Chapter 6
Conclusion

The importance of doubled haploidy is well known in all fields of agriculture and related disciplines. The acceleration has been observed in research studies on doubled haploid production over the last 5 years. Major research has been focused to change the status of many recalcitrant crops to responsive and to improve the overall methodology of doubled haploid production. Several model genotypes have been identified in various crops that have led to an overall improvement in the doubled haploid production technology. In this regard, IMC has been of special interest to the plant breeders, geneticists, and molecular biologist due to the availability of embryonic units in a larger number. It is quick and efficient. Moreover, genetically identical and physiologically uniform microspores provide a target for cell biology and genetic engineering studies. Molecular studies have led to an increase in our knowledge on the pathways by which gametophytic development is converted to sporophytic pathway during microspore embryogenesis. The genomic studies have identified various genes/QTLs associated with androgenic induction that will help for further improvement in the production of doubled haploids. Nevertheless, DH technology is a fascinating phenomenon and a powerful tool to speed up the breeding process for cultivar development and thus, can help achieve food security on a sustainable basis. Further work on early embryogenesis and especially the induction phase using video tracking systems and flow cytometry will definitely help address the challenges.

M. Asif, *Progress and Opportunities of Doubled Haploid Production*,
SpringerBriefs in Plant Science 6, DOI 10.1007/978-3-319-00732-8_6,
© Springer International Publishing Switzerland 2013

Index

M. Asif, *Progress and Opportunities of Doubled Haploid Production*,
SpringerBriefs in Plant Science 6, DOI 10.1007/978-3-319-00732-8,
© Springer International Publishing Switzerland 2013